CLIMATE JUSTICE AND HISTORICAL EMISSIONS

This volume investigates who can be considered responsible for historical emissions and their consequences, and how and why this should matter for the design of a just global climate policy. The authors discuss the underlying philosophical issues of responsibility for historical emissions, the unjust enrichment of the earlier developed nations, as well as questions of transitional justice. By bringing together a plurality of perspectives, in terms of both the theoretical understanding of the issues and the political perspectives on the problem, this book presents the remaining disagreements and controversies in the debate. Giving a systematic introduction to the debate on historical emissions and climate change, the book also provides an unbiased and authoritative guide for advanced students, researchers, and policy makers in climate change justice and governance and, more widely, for anyone interested in the broader issues of global justice.

LUKAS H. MEYER is a professor in the Department of Philosophy and Dean of the Faculty of Arts and Humanities at the University of Graz, Austria. Professor Meyer's research focuses on intergenerational justice, the ethics of climate change, and historical justice. He was leader of the research project *Climate Change and Justice: The Significance of Historical Emissions* from 2010 to 2014, and one of the lead authors of the Fifth Assessment Report of the Intergovernmental Panel on Climate Change (IPCC).

PRANAY SANKLECHA is an assistant professor in the Department of Philosophy at the University of Graz, Austria. Professor Sanklecha's teaching and research interests cover topics including the meaning of life, intergenerational justice, the ethics of climate change, methods of justification in normative philosophy, and philosophy as a way of life. He has written articles on intergenerational justice, particularly as it relates to climate change.

CLIMATE JUSTICE AND HISTORICAL EMISSIONS

Edited by

LUKAS H. MEYER
University of Graz, Austria

PRANAY SANKLECHA
University of Graz, Austria

CAMBRIDGE
UNIVERSITY PRESS

University Printing House, Cambridge CB2 8BS, United Kingdom

One Liberty Plaza, 20th Floor, New York, NY 10006, USA

477 Williamstown Road, Port Melbourne, VIC 3207, Australia

4843/24, 2nd Floor, Ansari Road, Daryaganj, Delhi – 110002, India

79 Anson Road, #06–04/06, Singapore 079906

Cambridge University Press is part of the University of Cambridge.

It furthers the University's mission by disseminating knowledge in the pursuit of education, learning, and research at the highest international levels of excellence.

www.cambridge.org
Information on this title: www.cambridge.org/9781107069534

First published 2017

A catalogue record for this publication is available from the British Library.

Library of Congress Cataloging-in-Publication Data
Names: Meyer, Lukas H., editor. | Sanklecha, Pranay, editor.
Title: Climate justice and historical emissions / edited by Lukas H. Meyer, Pranay Sanklecha, University of Graz, Austria.
Description: Cambridge, United Kingdom ; New York, NY : Cambridge University Press, 2017. | Includes bibliographical references and index.
Identifiers: LCCN 2016030025 | ISBN 9781107069534 (Hardback)
Subjects: LCSH: Environmental justice. | Transitional justice. | Climatic changes–Social aspects. | Environmental policy–Social aspects. | BISAC: LAW / Environmental.
Classification: LCC GE220 .C557 2017 | DDC 363.738/74–dc23 LC record available at https://lccn.loc.gov/2016030025

ISBN 978-1-107-06953-4 Hardback

Contents

Contributors

Christian Baatz is Faculty Member at the Department of Philosophy, University of Kiel, Germany.
E-mail: baatz@philsem.uni-kiel.de

Brian Berkey is Assistant Professor of Legal Studies and Business Ethics in the Wharton School at the University of Pennsylvania, an associated faculty member in the Penn Philosophy Department, and an affiliated faculty member of the Institute for Law and Philosophy at the Penn Law School.
E-mail: bberkey@wharton.upenn.edu.

Daniel Butt is Associate Professor of Political Theory at the University of Oxford; Fellow and Tutor in Politics at Balliol College, Oxford; and Director of the Centre for the Study of Social Justice, Oxford.
E-mail: daniel.butt@politics.ox.ac.uk

Daniel A. Farber is Sho Sato Professor of Law at the University of California, Berkeley. He is also the Co director of the Center for Law, Energy, and the Environment. He is a Member of the American Academy of Arts and Sciences and a Life Member of the American Law Institute.
E-mail: dfarber@law.berkeley.edu

David Heyd is Chaim Perelman Professor of Philosophy, Department of Philosophy, the Hebrew University.
E-mail: david.heyd@mail.huji.ac.il

Anja Karnein is Associate Professor of Philosophy at the Binghamton University, State University of New York, and has been a Visiting Fellow for the Center for Ethics, Harvard University; a Postdoctoral Research Fellow at the Center for Society

and Genetics at UCLA; and a Visiting Scholar at the New York University Center for Bioethics.
E-mail: akarnein@binghamton.edu

Sarah Kenehan is Associate Professor of Philosophy at Marywood University in Scranton, Pennsylvania (United States).
E-mail: skenehan@maryu.marywood.edu

Mizan R. Khan is Professor at the Department of Environmental Science and Management at North South University, Dhaka, Bangladesh. He has been attending the climate negotiations as a member of the Bangladesh delegation since 2001.
E-mail: mizan.khan@northsouth.edu

Lukas H. Meyer is Professor of Moral and Political Philosophy at the University of Graz, Austria.
E-mail: Lukas.meyer@uni-graz.at

Konrad Ott is, since 2012, Professor of Philosophy and Environmental Ethics at the University of Kiel, Germany.
E-mail: ott@philsem.uni-kiel.de

Pranay Sanklecha is Assistant Professor of Philosophy at the University of Graz, Austria.
E-mail: pranay.sanklecha@uni-graz.at

Rudolf Schuessler is Professor of Philosophy at the University of Bayreuth, Germany.
E-mail: rudolf.schuessler@uni-bayreuth.de

Janna Thompson is a Professorial Fellow at La Trobe University and Fellow of the Australian Academy of the Humanities and the Academy of the Social Sciences in Australia.
E-mail: J.Thompson@latrobe.edu.au

Introduction

On the Significance of Historical Emissions for Climate Ethics

LUKAS H. MEYER AND PRANAY SANKLECHA

Since industrialization human beings have contributed to what has become known as climate change. Human activities, and especially the production, trade, and consumption of goods, have greenhouse gas emissions as their side effect. Industrialized countries are responsible for more than three times as many emissions between 1850 and 2002 as developing countries (Baumert et al., 2005: 32). These emissions have long-term global effects on the median temperature on Earth with highly unequal and predominantly negative consequences for human welfare. So far the level of welfare realized in a country or region has been strongly correlated with the historical and current levels of emissions of this region. Accordingly, the efforts of developing and newly industrializing countries to "catch up" (and to close the huge welfare disparities between them and the highly industrialized countries) have gone together with an increase of total emissions (Khan, Chapter 10). Further, the increasing number of human beings on Earth will foreseeably cause higher total global emissions.

There is no easy technical solution. To reduce global emissions drastically will require major changes in how we produce, trade, and consume goods; to prevent harm from the consequences of climate change will require major adaptations and heavy investments. With business continuing as usual, very many people and mostly poor people living in the future will become victims of the consequences of climate change as they will not have the resources and capacities to adapt to dramatically changing environmental circumstances. Also, developing countries are more vulnerable to climate change as a result of geographical factors (e.g., already higher temperatures before climate change) and higher reliance on agriculture, which is an especially vulnerable sector. Levels of historical and current emissions are correlated both positively with levels of welfare realized and negatively with the levels of (especially future) harm suffered and basic rights violated as a result of the effects of climate change. In this way poor people have lost out both in terms of benefits realized owing to emission-generating activities and through these activities imposing the risks of rights violations predominantly on poor people living in the future.[1]

For the purpose of introducing the debate to which the authors of this volume contribute, we can distinguish between industrialized and developing countries in a highly simplifying manner that does not reflect the fact that many countries cannot be grouped purely in one or the other category. (For this idealized characterization of the problem of historical emissions, see Meyer and Roser, 2010 and 2006.) The distinction relies on the fact that there is a strong (though less than perfect) correlation between (i) having emitted more in the past, (ii) having more benefits grounded in past emissions, (iii) being less vulnerable to climate change, and (iv) being wealthier in general. All these features describe the countries that took part in the so-called Industrial Revolution compared to those that till today have not undergone industrialization.

The four features are correlated to a significant degree because of causal interdependencies. As the contributions to the volume illustrate, arguments that ascribe higher duties to some countries or people than others will be based on one of these four features (higher past emissions, higher benefits from past emissions, lower vulnerability, or higher wealth). So if those features all coincide in the industrialized countries, the dispute is not about ascribing higher duties to the industrialized countries. Rather, the task consists in correctly interpreting these reasons with respect to their significance and weight, thereby specifying why and to what extent industrialized countries can be assigned higher duties.

If one succeeds in doing this, one will at least have a basis for specifying the duties of those countries in which the four aforementioned features do not coincide. Indeed, these "impure" cases have become important and will become more important as a result of the catch-up economic developments in developing countries based on high-emission technologies and the fact that groups of people within countries differ considerably in terms of features (ii)–(iv). Thus the arguments developed by the contributors to the volume arguably could be applied not only to distinctions between the industrialized and developing countries and countries with different combinations of features (i)–(iv) but also to socio-economic groups within countries and to different individuals.

Relying on the highly simplifying distinction as introduced, the relations between industrialized countries and developing countries can be described as being doubly asymmetrical: First, the industrialized countries have the main

[1] The factual claims of this characterization are supported by the main findings of the recent assessment report of the International Panel on Climate Change. See, for example, *Synthesis Report: Summary for Policy Makers* at www.ipcc.ch/pdf/assessment-report/ar5/syr/AR5_SYR_FINAL_SPM.pdf.

For a review of the state of the art of the normative issues introduced here, see the relevant chapter of the assessment report: Kolstad et al. (2015).

historical and causal responsibility for climate change and have realized large benefits from emission-generating activities, whereas the developing countries have contributed far less to the problem of climate change and have derived comparatively small benefits from emission-generating activities. Second, for the reasons given, most importantly their far better adaptive capacities, industrialized countries are likely to suffer less damage and, in terms of human suffering and the violation of basic rights of persons, less severe damage from the resulting climate change. This finding suggests that, normatively speaking, the problem of how to respond to climate change can be understood as a distributive problem with a significant historical dimension and thus a problem of historical justice. However, it differs from other problems of historical justice and especially those that address cases of historical injustice in one very important respect. Unlike in standard cases of historical injustice, it is difficult to characterize the historical emission-generating activities of people as unjust, specifically before the problem of climate change became known and, thus, before these people could have known about their contributing to long-term global and harmful consequences. (For a discussion of historical injustice more generally, see Meyer, 2006; and the contributions in Meyer [ed.], 2004.)

Allowing for some variations in detail, the contributors to the volume can be said to make use of this idealized understanding of the problem. However, this does not by itself settle the question of how the historical dimension of the problem is to be assessed. It remains to be seen whether, and if so, how past emissions and their beneficial as well as harmful consequences are normatively significant for both the attribution of responsibilities and duties in responding to climate change and the distribution of the shares of the remaining permissible emissions. These issues are at the heart of the debate and the disagreements between the contributors to the volume.

There is further common ground among the contributors: They share the understanding that we need to distinguish between past (or historical) emissions depending on who caused them when. Mostly the contributors do not discuss the relevance of the past emissions currently living people have caused in their lifetimes. The amounts of these (and the accompanying benefits realized by people) differ dramatically, and arguably this should matter when it comes to both assigning fair shares of the remaining permissible emissions among currently living people and the duties they have to support those suffering harm and losses owing to the consequences of climate change. (For an argument to this effect, see Meyer and Roser, 2010.) The contributors also share the idea that today and for some time (most) agents engaging in emission-generating activities have or could have known about the long-term global consequences and that this matters for their duties with respect to climate change. In dispute are both the character and the normative relevance of the ignorance of previously living emitters and what the supposed or actual ignorance implies for the normative assessment of both their activities and their consequences.

To sum up, the authors of the volume focus mainly on one question of climate justice: How, if at all, and for what reasons, should the history of past people's highly unequal causal responsibilities for climate change matter for the distribution of the moral responsibilities to respond to the problem among currently living and future people? Philosophers and normative theorists have engaged in the discussion on the significance of historical emissions for approximately fifteen years (among the first publications in peer-reviewed philosophical journals are Meyer, 2004; Gosseries, 2004), and it can be considered a developed and sophisticated, but certainly not settled, debate. In the following, our aim is twofold. On the one hand, we will selectively describe some of the results of this ongoing discussion, and, on the other hand, we will indicate how the contributions to the volume engage with and advance that state of the art. That is to say, we attempt to describe the contributions to the volume in a way that places them in the context of some of the central normative–theoretical problems and debates connected with understanding the significance of historical emissions.

The Normative Significance of Past Emissions

When it comes to the assignment of responsibilities for past actions that had emissions as their side effect, the contributors to the volume disagree with respect to which principle should govern the distribution of these responsibilities. Further, they disagree about whether individuals or collectives are the appropriate bearers of responsibility for historical emissions. This reflects disagreements about fundamental theoretical issues in ethics. Here we will introduce only three of these underlying disputes: the question of whether the actions of past noncontemporaries can alter the moral duties of currently living individuals; the normative significance of the so-called Non-Identity Problem (NIP) in the context of historical harm-doing; and the relevance of the purported or actual ignorance of past people about the long-term consequences of their emission-generating activities.

When the authors disagree with respect to which principle should govern the distribution of responsibilities for historical emissions, they engage in an ongoing debate in which several distinctions have been introduced. (For a discussion of these principles, see, e.g., Meyer, 2013.) We can distinguish between principles depending on who stands under compensatory, restitutive, or redistributive duties to respond to the harmful and highly unequal consequences of past people's emission-generating activities. The Polluter Pays Principle (PPP) identifies bearers of compensatory payments for the harmful consequences of wrongful emission-generating activities of past people.[2] Whether these activities can be considered wrongful is under dispute, however. A strict liability principle and certain versions of the Beneficiary Pays Principle (BPP) identify bearers of restitutive payments for the wrongless harmful and unequal consequences of past people's activities. Other principles do not rely on establishing a causal or normative relation between past people's actions and the consequences of climate change for the welfare of current as well as future people. Regardless of the causal explanation for the highly differing abilities of current actors to respond to the highly unequal consequences of climate change, the Ability-to-Pay Principle (APP) identifies bearers of redistributive payments according to a principle of distributive justice that is meant to guide us in bringing about just outcomes in the future.

Further, independent of which principle they defend, the authors disagree over whether the bearers of responsibility should be considered individuals or collectives (especially states when understood to exist over many generations as transgenerational entities with a relevantly fixed identity).[3] Distinguishing between individuals and collectives as bearers of responsibility allows us to differentiate between individualistic and collectivistic versions of the PPP and BPP as well as the APP. (A collectivistic version of the PPP has been dubbed the Community Pays Principle or CPP.)

The contributors to the volume defend different versions of these principles and also differ on which agent they consider to be appropriately identified by those principles. Anja Karnein argues for an individualistic APP (Chapter 5). Mizan R. Khan supports a collectivistic APP as complementary to the PPP so that the PPP is qualified by a consideration of how well off states are. Both Khan (Chapter 10) and Thompson (Chapter 2) explicate the PPP in terms of collectives, namely, transgenerational states being the bearers of the duties (and, in doing so, they endorse what other theorists have discussed under the heading of the CPP). David Heyd (Chapter 1) explicates a principle of unjust enrichment as an interpretation of the BPP when applied to wrongless past activities with highly unequal consequences. According to Heyd industrial states are the beneficiaries of unjust enrichment, and this should be reflected in the distribution of duties to respond to climate change.

[2] The PPP could also be understood to identify bearers of restitutive payments owing to wrongless but harmful emission-generating activities. See later discussion and especially Section 2 for a discussion of (strict) liability for historical emissions.

[3] Or, if they do not disagree on this, the authors differ in their focus. Some concentrate on collective (state) responsibility, whereas others concentrate on the responsibility of individuals. Those who discuss the responsibilities of states can be understood to assume a two-stage (or multistage) process in which in a first stage climate justice between countries is determined, after which each country will internally distribute its mitigation and adaptation burdens fairly to individuals.

On the other hand, both Brian Berkey (Chapter 6) and Karnein (Chapter 5) discuss the question of currently living people standing under special obligations owing to the emission-generating activities of past noncontemporaries in terms of today's individual persons' moral individual responsibility. Karnein rejects the idea that the mere fact that currently living people have benefited from people's emission-generating activities could be considered normatively significant in imposing duties to help others in coping with the consequences of climate change. Rather, having benefited often goes hand in hand with being wealthy absolutely speaking and, thus, being able to help others, which Karnein identifies as the valid ground for "duties to help those most vulnerable to adapt to climate change" (Karnein, Chapter 5, 121).

Berkey argues that the attribution of duties to collectives (e.g., transgenerational states) and, in particular, the attribution of duties for past actions is incompatible with the notion of individual moral responsibility and should therefore be rejected. One way to understand the claim is the following: Fulfilling these duties will imply the authoritative imposition of costs (or restrictions) on individual current people who then stand under the duty to contribute their share (comply with the restrictions). Such imposition can be understood to reflect the idea that individual current people are morally responsible for past people's emissions. Berkey rejects such an understanding since moral responsibility of agents for certain actions or outcomes in his view presupposes that the agents can or could have acted differently or could have made a difference with respect to the outcome. Obviously, this is not the case for currently living people when it comes to past noncontemporaries' actions and their consequences. They could not possibly have had any impact on what their predecessors did as they had not yet been born. On similar grounds Karnein rejects the notion that currently living people can be considered as being implicated in wrongful action simply because of having benefited from the consequences of historical emissions.

Janna Thompson, however, proposes a different understanding of the duties of individuals qua members or citizens of transgenerational states (Chapter 2). The idea is that members of such states (to whichever generation they may belong) can be understood to share a general interest in their state's fulfilling its duties of justice and throughout the time of its existence. Arguably people can realize certain values only together and as members of structured and transgenerational groups; they can live under, say, secure conditions of justice or in a tolerant society only when they are members of a transgenerational well-ordered state (cf. Meyer, 1997; Thompson, 2012; Scheffler, 2013). For such a state to exist the institutions of the state and their actions need to live up to what (political) justice requires both within the state and in its relations to other entities. Given people's interest in living in a just state, and presupposing that people have a duty to each other to establish and secure conditions of justice (a notion often referred to as people's natural duty of justice (Rawls, 1971: 334–37)), individuals as members of states stand under duties to support their state in fulfilling its duties of justice. These duties are both forward- and backward-looking when we assume that harmful wrongdoing requires measures of compensatory justice. Accordingly, Thompson contends not only that past people's emission-generating activities may be assessed as harmful wrongdoing from a time-neutral perspective but also that the transgenerational state has been involved in these activities in such ways that this wrongdoing is to be understood as activities by people as members of the state. If so, the state can be understood to stand under duties of providing measures of compensation and its current members stand under duties to support their state in fulfilling these duties of justice.

The second underlying issue concerns the relevance of the so-called NIP (Parfit, 1984: 351–441, esp. 351–80; id; Parfit, 1986). The NIP is relevant when we discuss the question of whether individuals or collectives are to be understood as bearers of duties to respond to the consequences of past people's emission-generating activities. The NIP gives rise to doubts that currently living individuals can be understood to have been harmed or benefited by the consequences of past people's emission-generating activities. According to the common understanding of harm/benefit an action will harm/benefit a person only if the person is worse/better off than the person would have been had the action not been carried out. Arguably, past people's emission-generating activities are (very likely) among the necessary conditions of the coming into existence of today's individual persons. That is, had these activities not been carried out (or, to put it bluntly, without the industrialization in Europe and, with some delay, in other regions of the world), these very people would not exist because different people would have been born instead. However, having come into existence and as the person who they are is a presupposition for their realizing well-being today (and at a level, we assume, so that they or most of them cannot reasonably object to having been brought into existence). Thus currently living people cannot be said to have been made worse or better off by past people's emission-generating activities. And without having been made worse or better off, currently living persons cannot be understood to have their welfare rights being violated or to have been unjustly enriched. (For a discussion of alternative notions of harm doing and benefiting in the context of discussing the implications of the NIP, see Meyer, 2015: section 3.)

Owing to their interpretation of the implications of the NIP, both Heyd and Thompson argue in the volume in favor of a collectivist reading of the differing principles each of them endorses (versions of the BPP and the PPP, respectively). They both make the point that the NIP does not arise for certain collective agents in the context of past emission-generating activities: when we presuppose that these collectives have had a continued existence as one and the same agent spanning the relevant period, that is, the times when they can be considered responsible for past emission-generating activities and today, the past emission-generating activities cannot be understood to have had an identity-instituting impact. (See Page, 2006: 150–58, for an examination of this argument.) Just as the past emission-generating activities of currently living individuals do not affect their identities as the persons they are, historical emission-generating activities do not affect the identities of transgenerational states (and among them of the industrial or rich nation-states) so long as their identities can be considered fixed for the period including the time when these activities were carried out by them or in their name.

For Berkey, on the other hand, this way of responding to the NIP is irrelevant because, in his understanding and as introduced previously, the implications of attributing responsibility to these collective agents conflict with what can be legitimately asked of individual people as moral persons: not only could they not possibly have made a difference with respect to their past noncontemporaries' emission-generating actions, currently living individuals very likely owe their personal identities to what these past people did. Thompson's interpretation of people's individual responsibilities as sketched earlier can be understood as being immune to this criticism. Arguably for the attribution of responsibilities to people as citizens their personal identities are irrelevant, and so the NIP can be evaded. One could argue for this on the basis of Kumar's response to the NIP (Kumar, 2003; Kumar and Silver, 2004).

The third underlying issue concerns past polluters' epistemic state with respect to (their purported or actual) ignorance about the long-term and global consequences of their emission-generating activities, and to this we now turn.

The Relevance of Ignorance about the Harmful Effects of Emissions

In many cases, we seem to think that ignorance of the harmful consequences of an action is relevant to assessments of moral responsibility for performing that action: If an agent does not know the consequences of a given action, or does not know that the consequences are wrong, it seems unfair to blame him or her for performing that action. (For discussions of and differing views on this issue, see, e.g., Guerrero, 2007; Harman, 2011; Rosen, 2004; Zimmerman, 1997.) For example, let us say I flick a switch to turn off the lights in my office when I leave it. I do this every time I leave because I consider it wasteful not to do it. Unbeknownst to me, while I was out for lunch a malicious colleague connected the wiring in such a way that when I flick the switch, the university library blows up, causing death and injury to people and damage to valuable books. My flicking the switch is causally necessary for the deaths, injuries, and damage, but I would normally not be considered blameworthy for causing this harm (but presumably my colleague would be). (For a discussion of this example, see Duff, 2007: 75–77.)

However, it would be too hasty to conclude from this that ignorance always excludes blame. Consider another case: I may genuinely be ignorant of the fact that animals can feel pain. In part on this basis, I torture a baby seal. In this case, we would intuitively think that my ignorance does not excuse me from blame (and in this case no one else could be blamed). Rather than holding that ignorance always excludes blame, often a sort of hypothetical standard is being used; that is, regardless of what someone actually knows, we can identify what people can reasonably be expected to know. Being ignorant of what people can reasonably be expected to know is not, on this fairly common view, an exculpating condition.

A central issue, therefore, in assessing whether, how, and to what extent historical emissions ought to be taken into account when distributing the costs of responding to climate change amongst currently living people is the epistemic state of past polluters at the time they caused the emissions. How is this relevant for the assessment of their actions, the assessment of the consequences of these actions, and for how, if at all, these actions or their consequences alter the duties of current agents, be they individuals or transgenerational entities?

Past people, one may plausibly claim, simply did not know that their emissions would cause harm to future generations, and it was only at a certain point in time that people could have reasonably been expected to know that their emissions contributed to causing long-term harm globally.[4] If ignorance, and in the relevant cases reasonable ignorance, can act as an excusing condition, then past people cannot be held morally responsible for their emissions. And if this is the case, one may go on to argue that there is no original moral responsibility that currently living people can inherit from their predecessors (for an extended discussion of inherited moral responsibility, see Miller, 2007: 111–62), and so one might conclude that the highly unequal past emissions ought to be irrelevant when determining how to distribute emissions (or, more generally, the benefits and burdens of responding to climate change) amongst currently living people.

This is a plausible claim, and one held by quite a few of the contributors to the volume. As we have already seen, for example, Berkey rejects the idea that currently living people can be held responsible for past people's emissions. Karnein and Khan also argue that current agents cannot stand under duties of providing measures of compensation (on the basis of the BPP or PPP) for the consequences of actions that they both assess as wrongless owing to past

[4] The precise point is a matter of dispute. One popular date, which Sarah Kenehan, for example, uses on practical grounds, is 1990. But, as she also thinks, this is by no means necessarily the date that would be established after close analysis.

people's actual and excusable ignorance (while both hold that current agents stand under special obligations according to the APP or, this being Khan's explicit view, the APP as complementary to the PPP with respect to emissions since the date after which ignorance can no longer serve as excuse).[5] Rudolf Schuessler too, as we will see, bases part of his argument on what he calls "one of the most important claims in climate ethics," namely, "that the citizens of industrialized countries were *inculpably* ignorant about the greenhouse effect or the impact of greenhouse gas emissions on the earth's atmosphere during most of the emission histories of their countries" (Schuessler, Chapter 7, 157).

Heyd, however, rejects the move from holding (a) that past emitters were not morally culpable for their high levels of emissions owing to their legitimate ignorance of the consequences of those emissions to holding (b) that this means that historical emissions are irrelevant to determining how to distribute emissions going forward. In terms of the principles introduced previously, he proposes an interpretation of the BPP that does not require showing that the benefits accrued were produced unjustly. His contention is that the past polluters' ignorance is irrelevant for the causal relation between past people's actions and the present conditions of their descendants' being normatively relevant: it is sufficient that the descendants have been highly unequally benefited for past people's emission-generating activities creating responsibilities for their descendants in the present. That the beneficiaries are unjustly enriched is the consequence of past people's having made highly unequal use of what in the meantime we have learned to be a limited resource. Further, use of this limited resource has so far been (and will for the foreseeable future be) crucial to people's realizing welfare. The issue of the predecessors' supposed or actual ignorance can be put aside. Those enriched by their predecessors' use of a limited global commons stand under duties to provide restitution to those who have not benefited similarly from the use of the resource. For Heyd, it is important that current agents' responsibilities are responsibilities of liability and restitution rather than of compensation, because as it is commonly understood, the attribution of compensatory duties would necessarily reflect a moral charge of someone having done something wrong (Heyd, Chapter 1).

Thompson also thinks that historical emissions are relevant, but she differs in a key respect from Heyd: she argues that historical emissions were wrongful in a certain sense, namely, from a time-neutral perspective and, thus, that the

[5] Khan points out that even a generous determination of the date after which ignorance can no longer serve as excuse, namely, 1990, will attribute high proportional historical responsibility to the OECD countries of Europe (and the corresponding duties according to the PPP): "Moving the baseline year by a few decades does not dramatically shift levels of historical responsibility. For example, shifting the first year of counting emissions all the way from 1890 to 1990 decreases the contribution of OECD Europe from 14 to 11 per cent of the world total" (Khan, Chapter 10, 237). We will discuss this point later.

responsibilities of the collective agents are to be understood as compensatory in nature. Specifically, she argues for duties of industrial states as transgenerational entities to provide measures of compensation in response to their wrongful past high-emission-generating activities. According to Thompson past polluters' ignorance may excuse what they did at the time of their actions. However, their actions are nevertheless to be assessed as wrongdoing from a time-neutral perspective, as the epistemological lack is a failing that belongs to the agent (unless it could be shown to have been utterly impossible for the agent to have known about the consequences) and, thus, establishes nonculpable responsibility to repair the wrongs. As the historical high-emission-generating activities are to be understood as actions of the industrial state that has continued to exist, Thompson attributes the responsibilities to repair the wrongs to the industrial states (Thompson, Chapter 2).

In his contribution, Daniel Butt uses a similar strategy to Heyd, in that he also tries to separate the moral charge of someone having done something wrong from the question of whether highly unequal historical emissions may be relevant for the distribution of responsibilities for responding to climate change between current agents (Butt, Chapter 3). Following David Miller, Butt distinguishes between moral responsibility and remedial responsibility. Butt (and, as we saw, Berkey agrees with him on this) argues that currently living agents cannot be held morally responsible for the emissions of their ancestors, because "we cannot be morally responsible for outcomes over which we had no kind of input or control, and which in no way result from choices we have made" (Butt, Chapter 3, 62). However, currently living people may have remedial responsibility without having moral responsibility. As Butt describes it, "To bear remedial responsibility for a particular problematic outcome is to bear a special responsibility to correct the outcome, to act in such a way as to right whatever wrong or solve whatever problem has been created" (Butt, Chapter 3, 62).

Butt argues that currently living people can bear remedial responsibilities owing in part to the emissions of past people to whom they are connected. One way of rejecting this ascription of remedial responsibility is to appeal to the "exculpatory block" of the ignorance of past emitters. Butt argues that we can plausibly make a counterfactual claim here, namely, that even if past emitters had known about the harmful consequences, "they would have carried on regardless" (Butt, Chapter 3, 66). While making such a counterfactual claim does mean that "we make a moral appraisal of the character of the perpetrator in such a judgment," (Butt, Chapter 3, 70). it is crucial for Butt that this appraisal is done not with a view to assigning moral responsibility or punishment, but rather for determining remedial responsibility. The counterfactual claim does not allow us, and is not intended to allow us, to say that past emitters can be held morally responsible for the emissions about whose harmful consequences they were ignorant. Rather, what it does is allow us to assign remedial responsibility for those consequences to currently living people who are connected in the right way to the past emitters. The morally excusable ignorance of past emitters should not, given the counterfactual claim, be considered an exculpatory or "blocking" condition for assignations of remedial responsibility to currently living people.

Butt is careful in making clear that his argument against the exculpatory block of ignorance in the case of past emissions "does not amount to a general argument in favor of strict liability for the effects of emissions." (Butt, Chapter 3, 66). There are good dialectical reasons for this. As he discusses in his chapter, there is considerable philosophical objection to the idea of strict liability; for example, he quotes another of the contributors to the volume, Rudolf Schuessler, as suggesting "that adopting a principle of strict liability in a context where those who pay the costs were not aware of the effects of their actions 'comes close to primitive and/or totalitarian practices of clan liability' (Schuessler, 2011: 275)." (Butt, Chapter 3, 65).

Whatever the merits and strength of the philosophical objections to strict liability, it is a fact that legal regimes do use the idea of strict liability in certain contexts. In his contribution to the volume, Daniel A. Farber examines how this idea is used in some regimes and contexts, with the ultimate aim of exploring how this might carry over to the case of historical emissions (Chapter 4). Specifically, he examines the legal regimes of the United States, the European Union, and international law with respect to how they deal with environmental harms. His focus is on "analyzing legal settings where past acts can cause harm far in the future – product liability and hazardous waste – in order to examine how changes in technology and scientific knowledge affect liability" (Farber, Chapter 4, 81).

Farber himself notes three ways in which this analysis is useful to understanding the normative significance of historical emissions. First, "it could be relevant to considering the possible limits of legal liability for greenhouse gas emissions." (Farber, Chapter 4, 80). Second, "consideration of existing legal regimes might also be relevant to arguments about whether historic responsibility for past emissions could be implemented feasibly"[6] (Farber, Chapter 4, 80–81). Third, the "considered judgments of judges and government leaders in deliberating on analogous problems" (Farber, Chapter 4, 81). can contribute to ethical reflection, because they serve as some kind of data for ethicists who give intuitions and/or considered moral judgments a role in their ethical theorizing.[7]

[6] As we discuss Section 3, feasibility is another crucial issue in determining the relevance of historical emissions to the question of how to distribute emissions among currently living people.

[7] For "considered moral judgments," by "competent moral judges," see Rawls, 1951. Farber is careful to say that the considered judgments of judges and governments should not be understood *as* considered moral judgments per se, because the former are affected by various considerations that should be ruled out when making

Farber's analysis shows that while "liability based on fault seems to be the default rule," it is also the case that "stricter forms of liability are not infrequent, such as CERCLA in the United States and the Environmental Liability Directive in the European Union" (Farber, Chapter 4, 101). He points out that while strict liability regimes are often not absolute, in that they generally come with the possibility of a state-of-the-art defense, this defense is not to be found in all regimes and where it is found, can be interpreted differently. However, Farber also explains that for various reasons "international law has been particularly cautious about embracing strict liability for environmental harm" (Farber, Chapter 4, 101).

Farber's analysis and findings suggest at the very least that it is promising to assess the arguments for a strict liability regime in the context of historical emissions. Even if the current state of international law were not to allow for a strict liability regime for historical emissions and their consequences, Farber's analysis can help to broaden the range of normative interpretations we may want to consider. He points out that "it is not necessarily enough to simply distinguish negligence from strict liability". (Farber, Chapter 4, 101). That is to say, even if strict liability is not an option, this does not require going directly to a negligence regime. This seems related to the interpretations of Heyd, Butt, and Thompson, all of which can be understood to suggest that a straightforward assignment of moral fault to those who caused historical emissions is not necessary for assigning special responsibility to those current agents who are connected to the historical emitters and their actions in one way or another.

So far, we have seen that ignorance is discussed in the context of duties that relevant currently living people may or may not have to bear a greater share of the costs of responding to climate change owing to the emissions of past people. This is indeed the usual context in the debate as a whole. In his contribution, Rudolf Schuessler argues for a qualified and temporary reversal of one of the usual conclusions drawn in that debate. That is to say, instead of claiming that some currently living people should bear more costs owing to some relevant connection they have to past high emitters, he discusses the question of whether they should be given special consideration owing to how the emissions of past people have affected their ways of life (Chapter 7).

He places this argument in the context of the issue of grandfathering. Having considered and rejected extant defenses of grandfathering (Bovens, 2011; Knight, 2013 and 2014), Schuessler suggests what he calls a buffering principle: "People who seriously suffer under adaptive pressures through no fault of their own ought to receive the resources or the time they need to adapt in a morally acceptable way

considered moral judgments. The claim is only that they do serve as data in trying to figure out what the considered moral judgments of a given society may be.

if they lack the means to do so on their own." Acceptability implies that, if possible, the suffering of those people should not exceed a threshold that is to be determined by moral considerations" (Schuessler, Chapter 7, 153). (The costs of intervention may enter these considerations.)

He argues that this is a principle of justice, as "helping people digest undeserved adverse shocks and giving them time to adapt is one of the most plausible aims of luck-related ethics" (Schuessler, Chapter 7, 154). As we can see, for the buffering principle to apply, it is necessary that the shocks are not suffered as a result of immoral or otherwise blameworthy behavior (e.g., taking risks that would be considered excessive). He can then appeal to the ignorance of past emitters, that is, the claim "the citizens of industrialized countries were *inculpably* ignorant about the greenhouse effect or the impact of greenhouse gas emissions on the earth's atmosphere during most of the emission histories of their countries" (Schuessler, Chapter 7, 157) (On this issue, see Bell, 2011; Caney, 2010; Gardiner, 2011: 414–16; Schuessler, 2011.) Because, if they were indeed inculpably ignorant up to a certain point in time, then currently living people face adaptive pressures through no fault of their own and should therefore be given time to adapt to those pressures.

On this basis, Schuessler submits a novel interpretation of a claim that is relatively neglected in the debate. In his own words, he has "presented a luck-based moral argument for the temporary acceptance of proportional emission cuts (proportional to historically established emission levels at a reference point such as 1990)" (Schuessler, Chapter 7, 162) The buffering principle, along with the claim of inculpable ignorance up to a certain point, means that "the citizens of industrialized countries, who have to develop ways of sustainable living, are entitled to help, in particular by setting a suitable timeframe for proportional emission cuts," (Schuessler, Chapter 7, 162) and this not for pragmatic reasons but as a matter of justice.

This defense of temporary grandfathering is closely connected to what we have discussed under the heading of legitimate expectations of people living in highly industrialized countries when it comes to implementing a just climate change regime: it is important to consider the normative significance of the legitimate expectations that people in highly industrialized countries have with respect to being able to continue living in certain ways when thinking about how to transition to a just state of affairs with respect to climate change (Meyer and Sanklecha, 2014; Meyer and Sanklecha, 2011). We will say a little more about this later.

Responding to Climate Change Effectively and Ethically Defensibly

As we have seen, there are many issues over which one can disagree when it comes to whether, and if so how, historical emissions ought to be taken into account when deciding how we should respond to climate change today. Further, on each of those issues the disagreement consists of various possible positions rather than only two rival views. Like virtually all philosophical disagreement, none of this is likely to go away anytime soon. However, the disagreement must be practically overcome in some way, because – unlike with some other long-standing philosophical controversies – the subject of this disagreement is, ultimately, a problem of pressing practical urgency, namely, the problem of responding appropriately to climate change.

We need, in other words, to do something about climate change, and we need to do it quickly. Furthermore, some may think, the urgency of responding appropriately may mean that we should be willing to sacrifice some moral value if that is necessary for ensuring that we are able to respond effectively, and by doing so protect the basic rights of very many people. This takes us into the realm of ideal and non-ideal theory, an issue of central importance in theorizing about climate justice, and indeed about justice generally. (Important contributions to the general discussion are, among others, O'Neill, 1987; Mills, 2005; Sen, 2006; Farrelly, 2007; Miller, 2008; Stemplowska, 2008; Swift, 2008. For an overview, see Valentini, 2012, and our discussion in Meyer and Sanklecha, 2009: 16–26.)

To be clear, disagreement in itself is not necessarily a feature that makes a situation nonideal. But what is nonideal, amongst other things, is that we may not be in a situation to fulfill all elements of the relevant ideal. More concretely, we may be forced to choose between, on the one hand, reducing emissions enough so as to respond adequately to the forward-looking concerns of intergenerational justice and, on the other hand, ensuring that responsibility for past emissions is appropriately distributed.

This conflict is the central issue of Sarah Kenehan's contribution to the volume, where she assumes "that acting to prevent ... dangerous warming is the most important goal, the sort of goal that we might be willing to sacrifice other goods for in order to secure" (Kenehan, Chapter 9, 198). With this assumption in place, the focus of the inquiry changes – rather than attempting to describe what an ideal solution to climate change would be, the task becomes to outline a proposal that will prevent dangerous warming. Kenehan insists on the importance of both moral ideals and considerations of feasibility to how we should go about responding to climate change in the world as it is rather than as we might wish it to be. By paying heed to "the very real political hurdles and time constraints that face decision makers and world leaders," (Kenehan, Chapter 9, 198) she aims to argue for a proposal that manages to be both feasible, something that can actually be achieved, and is at the same time informed by moral considerations.

Kenehan's view reflects the understanding that we must not let the perfect be the enemy of the good when it comes to deciding how to take historical emissions into account: even if we accept that nations ought to take full responsibility for their historical emissions, this, she carefully argues on empirical grounds, is likely to be a political nonstarter. However, this does not mean that we should adopt a proposal of zero responsibility for historical emissions. Even if this were more feasible, it would run into the trap of "cynical realism," and consequently we should adopt a scheme of partial responsibility. Specifically, Kenehan argues on both practical and moral grounds, we should adopt a "proposal . . . that demands historical responsibility form 1990 on, as post-1990 emissions have a special moral character" (Kenehan, Chapter 9, 206)

How considerations of feasibility should matter for ethically defensible proposals of how to respond to climate change is an important and challenging question. The recent (recent in scholarly time) methodological debate in political philosophy about ideal and nonideal theory has led, among other things, to political philosophers' debating how, if at all, feasibility constrains what can count as morally valid recommendations for action. To schematize, at one end of the debate, there is the view that feasibility does not constrain moral judgments at all: justice is justice, regardless of whether it is possible. At the other end is the view that an ideal cannot be described as an ideal unless it is feasible. There is, as one might expect, much work on that connection, and on how to understand the relevant senses of feasibility. (See, e.g., Sen, 2006 and 2011; Farrelly, 2007; Cohen, 2009: 229–73; Gilabert, 2012; Gheaus, 2013.) What there is much less of, and what Kenehan provides, is a close and detailed analysis of feasibility in a particular context, and then an application of that analysis to the question how to choose among proposals that differ in feasibility and moral desirability. In this way, Kenehan contributes to the state of the art not just in the context of historical emissions and climate change, but also in the context of the debate on ideal and nonideal theory.

Christian Baatz and Konrad Ott also analyze the feasibility of certain proposals related to responding to climate change, and the relevance of feasibility to choosing among those proposals, but the question they tackle is different from Kenehan's. The focus of their chapter is a methodological question about how we should go about deciding what a just distribution of emissions is, an issue of high relevance to the question of how historical emissions should be taken into account when deciding how to respond to climate change (Ott and Baatz, Chapter 8).

Outside philosophical circles, one of the most popular proposals for the distribution of emissions is also the simplest: Every person is entitled to the same amount of greenhouse gas emissions. Within philosophical circles, however, this idea has begun to fall out of favor. One of the most important reasons for this is a pair of methodological criticisms made, amongst others, by Derek Bell (2008) and Simon Caney (2012). The criticisms are, respectively, that (a) it is a mistake to focus on emissions rather than climatic responsibilities generally (the charge of "atomism") and (b) it is a mistake to focus on the distribution of emissions in isolation from other considerations of justice (the charge of "isolationism"). And emissions egalitarianism makes both mistakes.

These criticisms have been made principally in the forward-looking context of how we should distribute emissions that must be capped in order to respond effectively to climate change. But it is worth noticing that, if they apply, they seem to do so also to the question of how historical emissions should be taken into account. For instance, if the so-called method of isolation is indeed a mistake, then the mistake seems to carry over to attempts to look exclusively at the relationship between historical emissions and the future just distribution of emissions, because this too focuses on emissions in isolation from other considerations of justice.

Theoretically speaking, the critiques have mostly been accepted. Practically speaking, however, there is a major problem: abandoning isolationism in particular risks making it virtually impossible to say anything about what the distribution of emissions should be, because it introduces so many other considerations as to make the question both theoretically and practically intractable (see Caney 2012 for a qualified rejection of this claim). Baatz and Ott argue that this intractability actually provides a good theoretical ground for continuing to be at least moderately isolationist in thinking about how to distribute emissions. Comparing their proposal with Caney's "modestly integrationist method of emissions allocation," (Baatz and Ott, Chapter 8, 186) they argue first that their own is significantly more feasible, and then that in this case feasibility is a virtue that matters. Furthermore, they make the interesting suggestion that we should not think of feasibility as an external constraint on ethical analysis; rather, in their view, "practicality/feasibility are an indispensable part of ethical all-things-considered judgments; they are not just an add-on to an already completed ethical analysis" (Baatz and Ott, Chapter 8, 188)

According to Baatz and Ott, feasibility is especially important "when dealing with transition or extrication ethics," (Baatz and Ott, Chapter 8, 188) This is, roughly speaking, the problem of how to move from nonideal to ideal circumstances, that is, from unjust to just circumstances. In our context, one of the questions of transition is: How do we move from an unjust distribution of emissions to a just one? As discussed earlier, Schuessler, in his chapter, argues that it may be necessary to institute a period of "temporary grandfathering," that is, a period when the higher historical emissions of industrialized countries are taken into account in order to give residents of those countries *more* emissions than they would receive under the fully just distribution of emissions (e.g., under emissions egalitarianism).

Schuessler's buffering principle, as introduced previously, is explicitly a principle of transitional justice, and it is important to note that he is not arguing that the eventual just distribution of emissions is one that will reflect grandfathering. Rather, the grandfathering is temporary, a method for arriving at, for example, emissions egalitarianism in a more just way than instituting it overnight. Further, he does not claim that the buffering principle overrides all other considerations. As he says, other "moral reasons ... may govern a transition," and consequently "the buffering principle has a *pro tanto* status" (Schuessler, Chapter 7, 154).

This is a way of framing a problem that we have discussed in our work (Meyer and Sanklecha, 2014; Meyer and Sanklecha, 2011). Citizens of industrialized countries live ways of life that are associated with a high level of emissions. We can assume that this is incompatible with the demands of climate justice. However, we must also note that abandoning or seriously revising those ways of life entails significant costs and, crucially, that citizens of those countries have over a long period been encouraged and supported by their states in choosing emission-intensive ways of life. They have an expectation that this support will continue, and the question is: What is the legitimacy of this expectation, and what should it mean for a just climate policy? Schuessler can be understood as arguing that the expectation is legitimate, and that what it means is that we should help people cope with the costs of the expectation's turning out to be frustrated, and that we should do so via temporary grandfathering.

But taking into account the costs that high emitters have in adapting to a way of life that is compatible with the demands of intergenerational and global justice is clearly only one consideration in assessing a transitional regime, namely, in terms of the fairness of its implementation. The main criterion should be that the transition effectively leads to a state of affairs that is both intergenerationally and globally just. Heyd can be understood to agree with Schuessler in arguing that we should specify principles governing the transition to conditions under which the application of ideal principles of distributive justice is possible. However, Heyd's main point here is that for this transition to work his unjust enrichment principle should be accepted. He argues for this also in terms of feasibility: without industrialized states' accepting their historical responsibility and temporarily making less than equal use of the limited resource of the adaptive capacity of the atmosphere, we will not be able to reach a just global distribution of benefits from emission-generating activities that, when it will have been reached, will allow the application of, for example, forward-looking (egalitarian) principles of distributive justice (Heyd, Chapter 1). Schuessler's temporary grandfathering or our notion of taking into account the legitimate expectations of people living in highly industrialized countries should be understood as a qualifying consideration: Other things being equal, a transitional regime that imposes fewer costs on people in adapting their ways of life to what the just climate regime requires is to be preferred (Meyer and Sanklecha, 2014: 384–89).

References

Baumert, K., Herzog, T., and Pershing, J. (2005). *Navigating the Numbers: Greenhouse Gas Data and International Climate Policy.* Washington, DC: World Resources Institute.

Bell, D. (2008). Carbon Justice? The Case against a Universal Right to Equal Carbon Emissions. In *Seeking Environmental Justice*, ed. S. Wilks. Amsterdam: Rodopi, pp. 239–57.

Bell, D. (2011). Global Climate Justice, Historic Emissions, and Excusable Ignorance. *Monist*, **94**(3), 391–411.

Bovens, L. (2011). A Lockean Defense of Grandfathering Emission Rights. In *The Ethics of Global Climate Change*, ed. D. Arnold. Cambridge: Cambridge University Press, pp. 124–44.

Caney, S. (2010). Climate Change and the Duties of the Advantaged. *Critical Review of International Social and Political Philosophy*, **13**(1), 203–28.

Caney, S. (2012). Just Emissions. *Philosophy and Public Affairs*, **40**(4):255–300.

Cohen, G. A. (2009). *Rescuing Justice and Equality*. Cambridge, MA: Harvard University Press.

Duff, R. A. (2007). *Answering for Crime: Responsibility and Liability in the Criminal Law*. London: Bloomsbury.

Farrelly, C. (2007). Justice in Ideal Theory: A Refutation. *Political Studies*, **55**(4), 844–64.

Gardiner, S. M. (2011). *A Perfect Moral Storm: The Ethical Tragedy of Climate Change*. New York: Oxford University Press.

Gheaus, A. (2013). The Feasibility Constraint on the Concept of Justice. *Philosophical Quarterly*, **63**(252), 445–64.

Gilabert, P. (2012). Comparative Assessments of Justice, Political Feasibility, and Ideal Theory. *Ethical Theory and Moral Practice*, **15**(1), 39–56.

Gosseries, A. (2004). Historical Emissions and Free-Riding. *Ethical Perspectives: Journal of the European Ethics Network*, **11**(1), 36–60.

Guerrero, A. A. (2007). Don't Know, Don't Kill: Moral Ignorance, Culpability, and Caution. *Philosophical Studies*, **136**(1), 59–97.

Harman, E. (2011). Does Moral Ignorance Exculpate? *Ratio*, **24**(4), 443–68.

Knight, C. (2013). What Is Grandfathering? *Environmental Politics*, **22**, 410–27.

Knight, C. (2014). Moderate Emissions Grandfathering. *Environmental Values*, **23**, 571–92.

Kolstad, C., Urama, K., Broome, J., Bruvoll, A., Fullerton, D., Gollier, C., Hahnemann, W. M., Hassan, R., Jotzo, F., Khan, M. R., Meyer, L., Mundaca, L., Olvera, C. (2015). Social, Economic, and Ethical Concepts and Methods. In *Climate Change 2014: Mitigation of Climate Change: Working Group III Contribution to the Fifth Assessment Report of the Intergovernmental Panel on Climate Change*, ed. O. Edenhofer et al. New York: Cambridge University Press, 207–282.

Kumar, R. (2003). Who Can Be Wronged? *Philosophy & Public Affairs*, **31**, 98–118.

Kumar, R., and Silver D. (2004). The Legacy of Injustice: Wronging the Future, Responsibility for the Past. In *Justice in Time: Responding to Historical Injustice*, ed. L. H. Meyer. Baden-Baden: Nomos, pp. 145–58.

Meyer, L. (1997). More than They Have a Right to: Future People and Our Future Oriented Projects. In *Contingent Future Persons: On the Ethics of Deciding Who Will Live, or Not, in the Future*, ed. N. Fotion and J. C. Heller. Dordrecht, Boston, and London: Kluwer Academic, pp. 137–56.

Meyer, L. (2004). Compensating Wrongless Historical Emissions of Greenhouse Gases. *Ethical Perspectives: Journal of the European Ethics Network*, **11**(1), 22–37.

Meyer, L. (ed.) (2004). *Justice in Time: Responding to Historical Injustice*. Baden-Baden: Nomos.

Meyer, L., (2006). Reparations and Symbolic Restitution. *Journal of Social Philosophy*, **37**(3), 406–22.

Meyer, L. (2013). Why Historical Emissions Should Count. *Chicago Journal of International Law*, **13**(2), 598–614.

Meyer, L. (2015). Intergenerational Justice. In *The Stanford Encyclopedia of Philosophy* (Fall 2015 Edition), ed. E. N. Zalta, URL: http://plato.stanford.edu/archives/fall2015/entries/justice-intergenerational/.

Meyer, L. and Roser, D. (2006). Distributive Justice and Climate Change: The Allocation of Emission Rights. *Analyse & Kritik* **28**(2), 223–49.

Meyer, L. and Roser, D. (2010). Climate Justice and Historical Emissions. *Critical Review of International Social and Political Philosophy*, **13**(1), 229–53.

Meyer, L. and Sanklecha, P. (2009). Introduction: Legitimacy, Justice and Public International Law: Three Perspectives on the Debate. In *Legitimacy, Justice and Public International Law*, ed. L. Meyer. Cambridge: Cambridge University Press, pp. 1–28.

Meyer, L. and Sanklecha, P. (2011). Individual Expectations and Climate Change. *Analyse & Kritik*, **32**(2), 449–71.

Meyer, L. and Sanklecha, P. (2014). How Legitimate Expectations Matter in Climate Justice, *Politics, Philosophy & Economics*, **13**(3), 369–93.

Miller, D. (2007). *National Responsibility and Global Justice*. New York: Oxford University Press.

Miller, D. (2008). Political Philosophy for Earthlings. In *Political Theory: Methods and Approaches*, ed. D. Leopold and M. Stears. Oxford: Oxford University Press, pp. 29–48.

Mills, C. W. (2005). "Ideal Theory" as Ideology. *Hypatia*, **20**(3), 165–83.

O'Neill, O. (1987). Abstraction, Idealization and Ideology in Ethics. *Royal Institute of Philosophy Lecture Series*, **22**, 55–69.

Page, E. (2006). *Climate Change, Justice and Future Generations*. Cheltenham, UK: Edward Elgar.

Parfit, D. (1984). *Reasons and Persons*. Oxford: Clarendon Press.

Parfit, D. (1986). Comments. *Ethics*, **96**, 832–72.

Rawls, J. (1951). Outline of a Decision Procedure for Ethics. *Philosophical Review*, **60**, 177–97.

Rawls, J. (1971). *A Theory of Justice*. Cambridge, MA: Harvard University Press.

Rosen, G. (2004). Skepticism about Moral Responsibility. *Philosophical Perspectives*, **18**, 295–313.

Scheffler, S. (2013). *Death and the Afterlife*. New York: Oxford University Press.

Schuessler, R. (2011). Climate Justice: A Question of Historic Responsibility? *Journal of Global Ethics*, **7**, 261–78.

Sen, A. (2006). What Do We Want from a Theory of Justice? *Journal of Philosophy*, **103**, 215–38.

Sen, A. (2011). *The Idea of Justice*. Cambridge, MA: Harvard University Press.

Stemplowska, Z. (2008). What's Ideal about Ideal Theory? *Social Theory and Practice*, **34**(3), 319–40.

Swift, A. (2008). The Value of Philosophy in Nonideal Circumstances. *Social Theory and Practice*, **34**, 363–87.

Valentini, L. (2012). Ideal vs. Non-Ideal Theory: A Conceptual Map. *Philosophy Compass*, **7**(9), 654–64.

Thompson, J. (2012). Identity and Obligation in a Transgenerational Polity. In *Intergenerational Justice*, ed. A. Gosseries and L. Meyer. Oxford: Oxford University Press, pp. 25–49.

Zimmerman, M. (1997). Moral Responsibility and Ignorance. *Ethics*, **107**, 410–26.

1

Climate Ethics, Affirmative Action, and Unjust Enrichment

DAVID HEYD

> As the partridge hath hatched eggs which she did not lay / So is he that hath gathered riches, and not by right.
>
> *Jeremiah 17: 11*

Many problems of inter-generational justice belong either to backward-looking historical issues (e.g., reparations) or to future-oriented matters (e.g., genetic intervention in the human genome). But there are some problems of justice in which past behavior and future conditions are inextricably connected. In these cases the present conditions are taken both as reflecting the wrongness of past behavior and as standards for the assessment of responsibility toward future people. The ethics of the distribution of the burdens of the mitigation of greenhouse gas emissions typically belongs to this third category, or so I wish to argue in this chapter. Just distribution of future emission quotas cannot be articulated without taking into account past behavior that gave rise to the present need for restraint.

But the particular way in which the backward- and forward-looking considerations should be combined in those issues of justice is highly contested. On the one hand, there is the example of poverty and the way it should be addressed. With all its a-historical ideal of equality, Marxist revolutionary theory partly argues for the dispossession of the privileged classes in historical terms, namely, as a kind of redress for long-lasting exploitation of the working class. On the other hand, it is often suggested that regardless of people's responsibility for their past behavior or the history of particular social and economic relations, justice should be applied to the present state as if it were a "clean slate." Accordingly, the provision of equal medical treatment even to people who have led an unhealthy life style is a typical case in point, but so is the exemption of children from responsibility for their parents' sins.

The way in which historical and future-oriented considerations should be combined in forming a just policy for the distribution of the burdens of climate change has been widely discussed. This chapter suggests that affirmative action can serve as an illuminating analogy exactly with respect to the tension between backward- and forward-looking forms of reasoning. It is surprising that this analogy has not received more philosophical attention, since the heated debate about the way to justify preferential treatment of minorities exhibits a structural similarity to the responsibility of the developed to the developing countries in preventing or mitigating the dangers of global warming. Indeed, it is true that historically the debate about reverse discrimination focused in the beginning on the moral call to compensate for past wrongs, whereas the environmental discourse on climate change two decades later originally concentrated on the call for present action for the sake of the future. But in both cases it has gradually become apparent that just policies cannot be ultimately shaped and justified without concern for both past and future circumstances.

However, the analogy teaches us that there is one major moral difference between the two kinds of causation of the present unacceptable state that requires our urgent response: racist policies and the institution of slavery were evil and their practitioners could and should have known that they were morally wrong (there were enough people in both North America and the rest of the world who vehemently opposed it on moral grounds). But the pollution of the atmosphere since the Industrial Revolution was not a moral evil and could not have been known to be wrong by those engaging in it, at least until a few decades ago. This casts a heavy doubt on the very possibility of ascribing moral responsibility to the past emitters of CO_2, let alone imposing duties on their current descendants. Despite the similarity between climate ethics and affirmative action in the complex blend of historical and a-historical considerations, the justification of the policy responding to them may be different. Thus, I wish to suggest that the way to bridge the gap between the "innocence" of the original polluters and the special historical duties of their descendants may be found in the concept of *unjust enrichment*. This originally legal concept will prove helpful in resolving the apparently mutually exclusive nature of the forward- and backward-looking approaches in climate ethics and maybe also in affirmative action. Like the analogy to affirmative action, the analysis of the historical responsibility for climate change in terms of unjust enrichment has been little discussed in the literature.

After outlining the background issue of the complex relations between the intra-generational and inter-generational duties in confronting global warming, the chapter will argue for a historically sensitive approach to climate change. Affirmative action will be introduced as an illustrative analogy for this approach. Then the main theoretical tool for justifying the historically based duties of the distribution of the burdens of mitigation, namely, the concept of unjust enrichment, will be outlined in some detail. Before concluding the discussion, a few empirical and theoretical difficulties in the analysis of climate ethics in terms of unjust enrichment will be considered.

Inter- and Intra-Generational Climate Justice

Here is a highly schematic, idealized (and empirically controversial) description of the background circumstances of the moral dilemmas in dealing with climate change (Figure 1.1.).

The starting point is some time in the past, let us say 1800, the beginning of mass emissions of CO_2 in the Industrial Revolution. We, in the present, are located somewhere at the middle of the time axis. The end of the curves can be imagined as a century from now. The upper curve describes roughly speaking the Western, industrialized, developed, and rich world; the lower one, the developing and underdeveloped poor parts of the world. The developed world is currently somewhere close to the peak of the upper curve. The developing world's peak lags behind and will occur sometime later. The distance between the two curves is not to be taken as referring to any quantifiable measure but only as expressing the constantly rising gap between the two up to the peak and the subsequent absolute decline of both – though the lower in a predictably much faster pace, unless we do something about it. Climate scientists predict a "tipping point," in which at a certain moment the

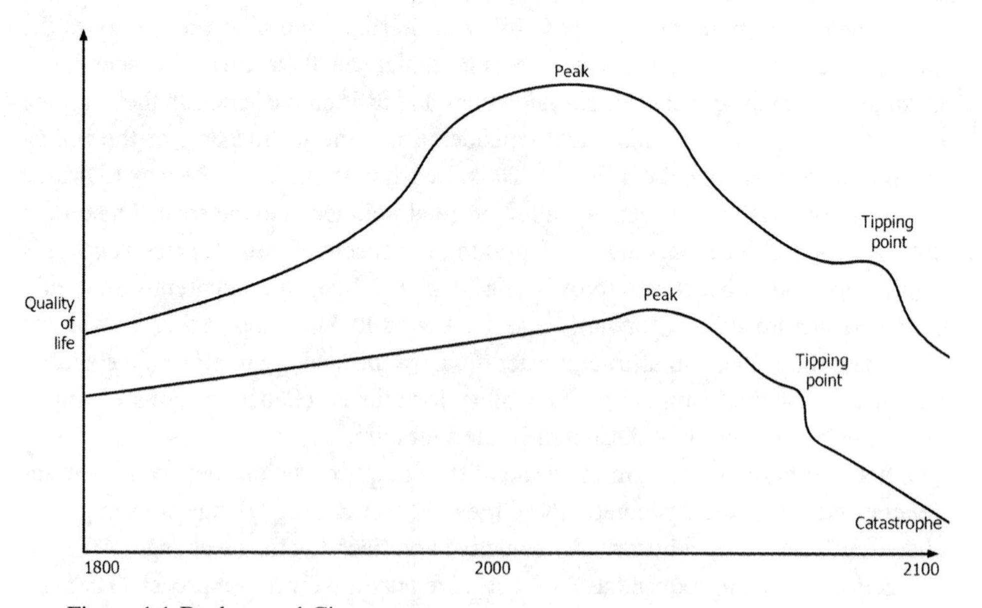

Figure 1.1 Background Circumstances.

mean temperature in a particular location will be hotter than the highest temperature reported in that location in the last 150 years (Mora et al., 2013). Tipping points refer to stages in the process in which the decline to catastrophic consequences becomes irreversible. Common scientific wisdom assesses that there is no way we can completely prevent the decline in both curves – at least for some period – but we do have control over both their *steepness* and the relative *distance* between them.

Present people are said to owe future people a real effort to mitigate the slope, and if possible to prevent it altogether. This applies primarily to those in the upper curve. They owe a natural duty of care to their own descendants but a duty of justice to the future people of the lower curve. The first duty is natural like that of a parent to their children and grandchildren. It does not depend on past conditions (or the shape of the curve). The other duty, that toward future people of the lower curve, is however of a different nature. It may have two sources: one relating to the *gap* between the curves; the other, involving the way in which that gap has been *created* and the way the circumstances of its formation are going to affect the condition of future people. The latter is the crucial dimension in the moral duty of present developed countries toward future people of underdeveloped countries: it is anchored in *historical* considerations. Furthermore, although the concern of industrialized nations for their descendants is a matter of pure inter-generational justice, the duties of those nations to future people in the world at large have significant *intra*-generational implications, namely, of present aid to the poorer nations. These are unsurprisingly the most difficult questions – morally and politically – since they require more than a natural concern for one's own descendants and raise the historically contentious question of responsibility for those past deeds that were ultimately the root of the imminent crisis.

Within a nation there is a way to compensate future generations for the damage of atmospheric pollution and the costs of restraining emissions. As John Broome has put it, there is a way to compensate future people for the damage of climate change "without sacrificing anything". We either leave them the side benefits of our pollution (in the form of the wealth generated by it) or reduce our current emissions and compensate ourselves by consuming more of other goods that do not involve further pollution, thus leaving our descendants less wealth but a cleaner environment (Broome, 2012: 8). However, I should note that this would not be a solution to the *global* problem of emissions since inheritance works on either the individual level or the nation's level. On the international level, where the historical inequality between nations was created, Broome's suggestion would do nothing to correct the injustice. It completely ignores the historical dimension and the crucial fact that much of the absorptive capacity of the planet has already been used by the first world countries and that any further use at the present rate would create catastrophic consequences for the whole world. This means that the past and present pollution of the atmosphere by the industrialized countries have been *at the expense of* the under-developed countries. And if that is the case, then the past exclusively inter-generational concern of each country for its own descendants has become inevitably an intra-generational issue in the current relations of the rich and the poor societies. The rich seem to *owe* the poor some assistance or compensation so as to enable them to fulfill their natural duties to *their* descendants.

I should emphasize that my argument stays within the general Rawlsian view on global justice. Countries do not owe as a duty of justice assistance to poorer ones just because they are poor. The difference principle does not apply on the global level (Rawls, 1999: 113–120). Each country leaves to its next generation(s) whatever it chooses to leave or whatever it has at its disposal as a matter of luck. In other words, the fact that some countries had the luck, the initiative, or the intellectual tradition that led to the Industrial Revolution and to the resulting material benefits, does not in itself impose a duty to share its fruits with countries that were "left behind." And there is nothing wrong in itself in the African and Asian countries joining the industrial stage at a later point in history. Had there been no limit in the absorptive capacity of the atmosphere these countries could have now developed their economies by using heavy industries that use cheap sources of (highly polluting) energy. But that is not the case. They joined the party more or less at the time when people realized that the absorptive capacity of the atmosphere is limited and is dangerously close to an intolerable level. Once the scarcity of this resource (namely, the absorptive capacity) is recognized, what is left of this resource must be justly distributed and the *historical* fact that some specific countries have directly brought it about becomes highly relevant. As we shall see, this redistributive obligation is not derived from some fault on their part but from the stupendous benefits they have amassed.

Historically Sensitive vs. A-Historical Approaches

Introducing historical considerations into debates about distributive justice always leads to even fiercer controversies, mutual recriminations, and ultimately deadlock. There is, accordingly, a natural temptation to leave history aside and address the distributive issue directly, as if starting from a clean slate (like, typically, in games). We have a pressing problem – global warming – and we must seek a way to cooperatively solve it by sharing the burdens and costs involved. Thus, we can decide together on the so-called cap of emissions (no mean feat in itself, since this involves some debatable normative factors)[1], and then distribute emission rights either on the basis of equality or on the basis of relative wealth of current societies or according to some other principle. Personal or collective carbon allowances may be tradable or non-tradable depending on the degree to which we wish to apply paternalistic restrictions on the way the poorer countries with the larger allowances use their emission rights. However, in real social life the slate is never clean. Fixing the baseline for a just distribution in the present is morally acceptable only in ideal circumstances, that is to say, when the present distribution is not a consequence of past unjust behavior. Both Nozick and Rawls recognize this historically sensitive condition of justice and leave ample room for corrective justice before the application of their general principles of distribution. Although Rawls offers a "patterned," a-historical theory of justice, he admits that it must be complemented by a historically based theory of compensation. And although Nozick seems to sanctify the present distribution irrespective of any a priori pattern, this applies only to cases in which no injustice of acquisition or transfer occurred in the past.

So it is only natural that poorer countries today make claims against the richer ones in historical terms, claiming that to take the present as a baseline for the distribution of the burdens of climate change would be unfair to them. But do they have a case?[2] Climate change is more problematic than standard cases of historical wrongs that call for compensation. Those who caused the current crisis are dead. Even if they were alive, they could not be said to have wronged future people since they did not know and could not know that the emission of CO_2 was harmful. And even if they knew it for fact, we would not be able to identify them personally. And even if we could identify them, the Non-Identity Problem (NIP) would block us from making claims against them, since it is likely that we would not have been born had the alleged wrong not been done. Harm can exist only where there is an identifiable perpetrator, an identifiable victim and the right causal relation between the two (Heyd, 2009). These conditions are not satisfied in the case of global warming.

There is thus a tension between the natural inclination to treat the ethics of climate at least partly in historical terms and the conceptual and moral difficulties in supporting this inclination theoretically. For example, there are theorists who believe that the "ability to pay" is the right guiding principle for the distribution of the burdens of fighting global warming, requiring mass transfers of capital from rich to poor countries. Even if the gap between the rich and the poor is *partly*

[1] Thus, when we refer to possible "catastrophic" consequences of global warming we have to agree, for example, whether we mean by that the dramatic reduction of car and airplane travel with all the far-reaching economic effects on our standard of living, or whether we mean the actual physical damage to our environment by massive flooding or by the spread of life-threatening epidemics.

[2] See, for example, Stephen M. Gardiner (Gardiner, 2010: 14–16).

the result of differential polluting practices in the past, this climate component is irrelevant to the duty of justice of the former to the latter since the wealth of the developed countries already consists of the benefits of past industrialization with the pollution involved in it (Margalioth, 2012: 58).[3] The contingent fact that the rich countries in general are also the historical emitters of greenhouse gasses makes it easier to avoid judgment as to whether the source of the duty of justice is – in the language the United Nations Framework Convention on Climate Change (UNFCCC) – the differential (historical) responsibility for the current condition or the (a-historical) gap in the capabilities of the rich and the poor countries to address this condition now. This is the attraction of the "clean-slate" argument.

But this is misleading. The theoretical way by which we justify the heavier burden on the developed countries is important. For example, once we try to *fine-tune* the burdens imposed on the rich countries we realize that there are countries that are rich and have the ability to pay but which can hardly be ascribed with historical responsibility for global warming. Think of Switzerland that has water as a major source of energy or of Japan and South Korea that have used fossil energy in large quantities only since World War II. Should they have the same responsibility as the United Kingdom and the United States? Imagine that people became aware of the dangers of gas emissions in 1945: would not Japan and South Korea have argued then that it was unfair to require them to restrain their efforts of industrialization and modernization? We could argue (and will do so below) that even those "late comers" to the industrial world have actually benefited much from the products and technology of the older polluters. But imagine a country that completely independently of the achievements of the Industrial Revolution created green energy and reached a high level of production and wealth (say, a Switzerland with no historical connection to the European and American development of industry): would such a country be called to contribute to the global effort of mitigation? This would intuitively look unfair and hence can serve to highlight the ethical relevance of the historical principle.

The a-historical approach distinguishes between mitigation and adaptation costs. Mitigation is to be undertaken according to an egalitarian principle: everybody or every country should be now making the same effort – either by assigning tradable equal emission quotas or by imposing a carbon tax on emissions. If developing

[3] Margalioth argues that the Ability-to-Pay Principle should be evaluated according to the country's *overall* wealth rather than just according to the specific costs it incurs as a result of climate change. This makes sense as an a-historical principle of distributive justice, but it ignores the historically based damage imposed on poor countries independently of their current relative wealth. Accordingly components of wealth, such as pollution and colonialism (in contrast to such factors as Protestant ethics or scientific tradition), create historical responsibility (rather than obligations based on pure Rawlsian, a-historical distributive principles of justice). In many respects, despite critical misgivings, Margalioth, as a "welfarist," follows Eric A. Posner and David Weisbach (Posner and Weisbach, 2010). See also "Analysis of the US Case in Climate Change Negotiations" (Margalioth, 2013).

countries are given more emission allowances, this could give them an unfair edge in the global competition for industrial goods that would be cheaper for them to produce. Adaptation costs, on the other hand, may be seen as a matter of distributive justice that imposes duties of assistance of the rich to the poor (e.g., if their territory is flooded or their population is afflicted by epidemics) (Margalioth, 2012: 60–63). However, I don't find this distinction compelling since one can argue that *both* mitigation and adaptation should take into consideration the historical causes of both the dangerous depletion of the global public good of clean air and the specific grievous harms to particular countries like flooding. Thus, Bangladesh can claim special emission allowances as well as direct help in building ramparts because it was doubly harmed – first by being deprived of its *opportunity* (now) to develop its industrial production cheaply and second by being directly *harmed* (possibly in the near future) by the rising sea level due to its low topography.

Affirmative Action

An interestingly similar tension between historical and a-historical approaches to climate ethics can be found in the debate about affirmative action, an issue that preceded that of climate ethics by roughly two decades. This analogy, which has been hardly considered in discussions of climate change, highlights the specific problem of addressing an undesirable or even dangerous present condition that may negatively affect future people but is a consequence of actions performed in the distant past.[4] But historically the direction of the debate in the one was the opposite of that in the other. While the public debate on global warming started with the realization that there is a current global crisis and that we should be thinking about saving future generations and the planet from a catastrophe, affirmative action began as a backward-looking social project of compensating for the disgraceful discrimination of minorities (primarily African Americans) in the past. But once the debate on how to distribute the burdens of mitigating global warming heated up, historical claims began to be raised. In a parallel manner, soon after policies of affirmative action were challenged in the courts, the case for them changed and became more forward looking. Thus, following the famous *Bakke* case in 1978, universities have started to ground their preferential hiring and admissions policy in future-oriented arguments; for the courts permitted deviation from a pure color-blind policy only when diversity in class was the value to be

[4] One exception is Axel Gosseries: "While one might believe that policies such as affirmative action would better be justified on the grounds of distributive justice rather than on the reparation-for-injustice grounds, it is a striking feature of public debate that the latter rationale keeps resurfacing" (Gosseries, 2004: 55). Gosseries does not elaborate the analogy, but he is certainly correct in pointing to the persistence of the appeal to historical arguments in the political debate about both race and climate issues.

promoted. The interpretation of the Constitution by the courts blocked the historical argument for inverse discrimination primarily because it was considered incompatible with the individual rights of present people. Bakke could not be shown to have wronged any minority student and could not be himself discriminated against for belonging to the white majority that in the past collectively wronged the black minority. Consequently, the general argument for affirmative action moved to a-historical reasoning: the need for role models in minority groups, the importance of breaking racial stereotypes, the creation of equal opportunity to African Americans who come from lower-quality schools, and the intrinsic value of cultural diversity.

Ronald Dworkin, one of the most prominent advocates of affirmative action, is probably inclined to the a-historical approach due to his general commitment to the American Constitution and to the prominent role of individual rights in the evaluation of social policies. However, he argues that no individual has either a general right to be admitted to law- or medical school or even a specific right to be admitted on the basis of the exclusive criterion of intelligence (or SAT results). Social institutions may also use other criteria and those include racial diversity in some circumstances or the future benefit for society as a whole. In that respect, Dworkin staunchly opposes the idea of color-blindness in admissions policies (often promoted by the courts and by conservative opponents to affirmative action). He proposes that "in certain circumstances a policy which puts many individuals at a disadvantage is nevertheless justified because it makes the community as a whole better off" (Dworkin, 1977: 232). And he understands "better off" in either utilitarian terms (social welfare) or in ideal terms (social equality). But both are looking to the future.[5] One might wonder why if these social goals justify harming (innocent) individuals they cannot include also backward-looking reasons, such as past racial discrimination, which is the major cause for the current need for future-directed social policies. In other words, Dworkin could have easily introduced the historical factor of slavery into his theory of affirmative action without weakening his commitment to individual human rights of present people. Since he does not do so, there is a risk that the unique responsibility of current America to its African American minority would be no different from that responsibility to its Hispanic minority (for whom affirmative action may also create more welfare and more equality in future American society). A more just society is not only a more equal society; it is also a society that compensates members for past wrongs done to them that are partly responsible for the current inequality.

[5] See *Sovereign Virtue*, chaps. 11 and 12 (Dworkin, 2000). "Universities do not use race-sensitive admission standards to compensate either individuals or groups: affirmative action is a forward-looking, not back-ward looking, enterprise" (ibid.: 424).

The alternative line of argument according to which history is a major component in the justification of affirmative action seems more compelling. Take for instance Anthony Appiah's claim that affirmative action can be justified with no appeal at all to the language of rights, or alternatively by resorting to the idea of group rights. According to Appiah, the *remedial* function of affirmative action does not lie in "the fact that its agent has done some wrong but rather [in] the fact that some group has been wrongly disadvantaged" (Appiah, 2011: 272). There is no way to show on an individualistic theory of affirmative action how a present person x was harmed by a past person y. But that is not necessary since y and his descendants have benefited from the racial discrimination of x and other members of his group in the past (as well as in the present). My point is that beyond the general controversy on group or collective rights, racial discrimination is typically *systemic*, that is, characterizing a whole social system of benefits, privileges, status, and profit – all of which are often passed on from one generation to the next. The state (or if you wish, nation, society, economy) is by its very nature a trans-generational entity as are some of its constituent groups (or classes, minorities, genders). Thus, unlike individuals, whose lives last for a definite limited time, states are responsible for their longer-term past harmful deeds, and groups suffering from such deeds deserve remedies.

The analogy between climate change and affirmative action can be obviously challenged by noting the crucial difference between the two – the knowledge or awareness of the wrongdoer of the morally dubious nature of his actions. Industrial developers had no idea whatsoever of the damage they did to the environment and to future generations. The slave owners – even those who believed that slavery was morally justified – knew that the practice was morally controversial. In that respect, they may be held responsible for their deeds even if these were based on sincerely held beliefs and worldview. Slave owners *should have known* that they were engaged in vicious exploitation. We cannot ascribe similar knowledge to industrialists using coal in 1780. (Some even claim till 1980!) Accordingly, philosophers like David Miller, who takes the standard of knowledge as the criterion for the kind of justice that should guide policy, argue that contrary to affirmative action, which is in principle past-oriented, the burdens of reducing pollution should be anchored in future-directed principles like giving "sufficient" (rather than equal) opportunity to third world countries to develop in the current global circumstances (Miller, 2008: 136). Contrary to Miller, I propose that the standard of awareness does not affect the relevance of the past wrongdoing to the present issue of the just distribution of mitigation burdens and that the analogy of affirmative action and climate ethics still holds. As will be argued in the next section, all that is needed for taking the past as creating responsibilities for certain people in the present is the actual *causal* relation between the actions of people in the past and the present conditions of their descendants. No awareness of the wrong at the time of its commitment is required.

The climate change crisis can thus be considered as calling for some sort of inverse discrimination. Before we get to the stage of desirable color-blindness sometime in the future we have to give those who have been victims of unfair treatment some compensation and create equal opportunities for them, even in ways that seem to go against the ideal of racial blindness. Similarly, before we give nations or states equal emission allowances for further developing their production and consumer potential, we must give the under-developed nations permits to bring their industries to some proper baseline from which they can fairly compete with other countries using equal allowances. This would also involve some actions that go against the ideal of reducing greenhouse gasses, that is to say, the use of traditional, cheap but polluting sources of energy. In both cases inverse discrimination serves as an *interim* policy, which might be seen as incompatible with the ideal to which we strive (equal reduced emissions quotas or color-blindness) but which settles the moral account for damages created in the past.

Furthermore, in both racial relations and the issue of climate there is something arbitrary in imposing the responsibility for past damages or wrongs exclusively on the current generation. Therefore, the process of compensation and the creation of equal opportunity in both cases may take some time and should accordingly be distributed between us and some future generations. This is a typically long-term project since both racial discrimination and atmosphere pollution are still continuing in our generation despite the social awareness of the need to curb them. Even the consideration of creating "role models," which is often quoted as justifying affirmative action, has its counterpart in climate ethics: although practices such as avoiding the use of plastic bags in supermarkets or investing extra money in buying a hybrid car do not have a major effect on the level of CO_2 in the atmosphere, they do create an ethos of awareness of the climate crisis and a political consciousness that might put pressure for a more significant change in the way we, and particularly our governments and industries, manage energy consumption.

The NIP is an embarrassing challenge to the historical approach in dealing with current racial gaps and the climate crisis. For it is not only the question why should *I* pay for my ancestors' sins and why should the descendant of the victim of these sins be compensated when she is doing quite well now; it is the question whether I (as well as the compensated party) would have existed at all had the wrongdoing not taken place.[6] The NIP applies also to future people whose existence may be

[6] Rahul Kumar has challenged me with a most interesting argument: the NIP applies not only to people but also to the substantive *issues* or goods: thus, the harm done to the present generation in third-world countries cannot be defined independently of the Industrial Revolution itself. In other words, the good these people lost (the use of

dependent on the policies that we now pursue. But, as I have already noted, it is not unreasonable to assume that in some moral and political issues, like affirmative action, we may assume the existence of group rights and collective responsibilities. Global warming is a typical case in point and so is the treatment of the acute social gap in the status of white and black Americans. Once we adopt the view that entities such as states or nations maintain their existence and identity beyond the existence and identity of particular individuals comprising them, we can simply ignore the NIP. States can be taken as responsible for both past harms as well as future welfare.[7] They have interests (and some would add even rights, like self-determination). This is a common assumption in political philosophy, although I cannot defend it here. In any case, the problem of non-identity is logical; that of the status of states or societies – metaphysical.

But group rights, and especially responsibilities, are often in conflict with either individual rights or with rights of *sub*-groups who shun the responsibility of the main group. Again, there is an interesting parallel between affirmative action and climate ethics. Jews and Hispanics in the United States argue that not only did they not take active part in the practice of slavery; they were not even in the country at the pre-abolition period. The Japanese, the Norwegians, or the Swiss can similarly claim that they were only marginally engaged in the history-long pollution of the atmosphere because they were either late-comers to the Industrial Revolution or enjoyed alternative clean sources of energy. Thus, these nations do not wish to be included in the larger group of "The West" or "The Developed Countries" for that matter, and having had a different history they should be exempted from the collective responsibility of the larger group. This demand for exemption makes more sense in the case of whole countries than in that of sectors within a particular country. For the welfare – even identity – of the sub-groups in the case of American Jews or Hispanics is much closer to the identity of the group (i.e., the United States) than in the case of specific rich countries being part of "The Industrial West." Still, I want to argue that in both cases the responsibility of the sub-groups is partly grounded in the *benefit* that they derived from being associated or part of the larger respective group. White immigrants to America enjoyed the

the absorptive capacity of the atmosphere) was itself created by industrialization, that was the cause of their loss. One may try to address this challenge by the analogy to land in Locke's theory: although we do not know all the possible uses of land, and although people cannot be blamed for its over-use as long as they cannot recognize it as over-use, their enrichment by such over-use may require restitution later on. The air around us, no less than the land on which we live, is a universal public good that should be fairly shared also on the trans-generational level and irrespective of the *actual* awareness of what "leaving enough" for others amounts to. See footnote 21.

[7] This is a valid argument only if "states" can be identified independently of the harmful process of industrialization. Pranay Sanklecha has suggested to me the possible hypothesis that the nation state itself is a product of the Industrial Revolution. I am not sure this is empirically or historically true, but I admit that if it were, it would have consituted a serious conceptual obstacle to my argument – again, as a result of the NIP.

privileges and social benefits from which African Americans were excluded, including the historical benefits of past exploitation and oppression. Japan and South Korea (fortunately for them) joined the Industrial Revolution with its historically based advantages before humanity became aware of the limited absorptive capacity of the atmosphere, and they made full use of that opportunity that is not anymore open to those who are just now building modern economies.

But as long as we insist that global climate justice should be based not only on purely distributive principles but also on historical responsibility, the question remains: How can individuals, institutions or states be held responsible for compensating victims of past wrongs or damages if the perpetrators were not even aware of the danger of their actions, let alone knew them to be wrong? In that respect, climate ethics is different from typical cases of historical justice such as colonialism or unjust war. This difference might push us back to a "clean-slate" approach. However, I believe that we can explain and justify that persistent "resurfacing" of the historical element in the climate debate (as well as in the racial debate).

Unjust Enrichment

There is some talk in the literature on climate that contemporary people in the developed world are the undeserving beneficiaries of past actions of their ancestors.[8] But there are only a few attempts to conceptualize this "benefit" in terms of the legal concept of *unjust enrichment*.[9] Importing this concept from civil law on the level of individuals to global justice on the state or national level faces some obstacles, but it nevertheless offers a potentially promising way of accounting for the historical injustice involved in the clean-slate approach. It may offer an alternative to the idea of providing equal pollution permits to states or individuals as if the history of the present atmospheric condition may be ignored.

The verse from the book of Jeremiah quoted in the beginning of the chapter is enigmatic: "As the partridge hath hatched eggs which she did not lay: so is he that hath gathered riches, and not by right." This is the first source of the concept of

[8] One example is Lukas H. Meyer, who correctly states that neither our ancestors' ignorance, nor the fact that we are not responsible for their past deeds, nor the NIP can undermine the validity of the "beneficiary theory," which imposes on the developed world special obligations of mitigation of emissions (Meyer and Roser, 2006: 241-2).

[9] A thorough analysis of the advantages of a "beneficiary-pays-principle" in terms of non-wrongful historical harms is offered by Edward A. Page (2012). I fully accept Page's distinction between the wrongful and unjust categories of enrichment and the superiority of the latter as a model for climate justice, but wish to add a possible analysis of what *makes* past deeds from which present people benefit unjust. For many actions and policies of our ancestors may turn out to be (unbeknownst to them) beneficial to us and even at a cost to others, yet not necessarily be unjust. Consider a country that was not aware that it is over-fishing the oceans to such an extent that at a certain stage *all* countries will have to restrain their fishing practices in order to save some fish for future generations. It does not seem that the fishing country should carry all or even most of the burden of restraint despite having benefited for a long time from the fish. This is the force of applying a Lockean proviso according to which clean atmosphere may be compared to land, but not to fish. See note 21.

unjust enrichment. It refers to a bird that hatches eggs laid by other birds. The verse takes it as a simile to those who gather riches in an unjust way. The Hebrew Bible does not elaborate the meaning of "not by right," but it seems to imply that the wrong perpetrated by the partridge is outright theft.

However, later Jewish commentators, in the Talmud, have a more specific and sophisticated analysis of the concept of unjust enrichment. They are primarily concerned with cases in which the damage caused involves no fault. For example, if a goat wandering around in a public location consumes someone's barley from a sack, should the owner of the goat be held liable to the owner of the barley? Although there was no fault (e.g., negligence) on the part of the goat owner (the goat did not enter the premises of the barley owner), the goat owner must compensate the barley owner, because the goat (and its owner) *benefited* from the event. However, if the food consumed by the goat was bad for its health ("unwholesome"), the goat owner does not owe anything to the barley owner since there was, on the one hand, no fault on his part and, on the other hand, he did not benefit from the occasion. Such cases point to what we call "strict liability," that is to say, responsibility for events that involved damage but no bad intention, awareness of the agent, break of contract, or even negligence.[10]

Here the Talmud makes a crucial distinction: the case in which "the defendant derived a benefit and the plaintiff sustained a loss" as against the case in which "the defendant derived a benefit but the plaintiff sustained no loss." The case of the goat and the barley in which the plaintiff lost his barley is accordingly contrasted to a case in which one person occupies the premises of his neighbor which the neighbor did not intend to let (or could not let). The squatter can argue that although he benefited from occupying the premises, the owner sustained no loss and consequently is not entitled to compensation.[11] (Is not that exactly the moral argument of poor squatters of unoccupied houses in our big cities?) The Talmudic distinction is directly applicable to the history of pollution. It is indeed true that in the first stage of industrialization, in which no one could foresee the *global* effect of pollution, the industrial country could make the claim that it benefited from the intensive emission of CO_2 but that other countries did not lose anything (if there was any harm in pollution it was to the local people who lived close to factories emitting heavy smoke). But in retrospect we know now that the benefit of the polluting societies ultimately created significant externalities that are now imposed on the developing world.

[10] I prefer not to characterize unjust enrichment as a kind of strict liability in the trans-generational context of climate change, since strict liability cannot be ascribed to dead people, and we – the living – cannot be strictly liable for the cumulative scope of global warming because strict liability presumes an agent who at least in principle could have prevented the damage.

[11] See a good analysis of the squatter's case in *Unjust Enrichment* (Dagan, 1997: 112–120).

The Talmudic distinction may be used to show why the history of pollution is more than just a case of *free riding,* as Axel Gosseries argues (Gosseries, 2004). In free riding, one person takes advantage of the benefits that others have accrued by cooperating in their creation, as in Nozick's example of some public service for which all residents have paid except the free rider (Nozick, 1974: 90–95). The residents cannot be said to have been "harmed" by the person who was not willing to pay "his share." The free riding is certainly unfair but the residents cannot claim compensation for a *damage* they suffered from the behavior of the free rider. The history of global warming in that respect demonstrates a morally more serious situation: the industrial countries did not take a free ride on the non-industrial ones: they actually caused harm to the latter. They were unjustly enriched not just by not paying their fair share (of avoiding an environmental global hazard) in the past but by actually polluting the environment of the non-industrial countries and indirectly blocking their opportunity to develop cheaply generated industries in the present. And unlike the later generations of the polluting countries that have inherited the wealth amassed by their ancestors, the current generation of poor countries has not enjoyed any such inheritance. In other words, a free rider may turn out not to profit or even (ultimately) to lose by his free riding. His sin is being unfair in reaping benefits for which he did not pay. Being unjustly enriched, in contrast, necessarily involves some real loss to others (like the bird whose eggs are stolen by the partridge). At least in the strong sense that is relevant to climate ethics, unjust enrichment is not a case in which one party benefits while the other sustains no loss (as in the case of the squatter who occupies an "un-rentable" house), but rather more like an invader who occupies property that the owner could and would actually let. It is also the case that the free rider is aware of being a parasite while this is not the case for the unjustly enriched. One last difference between free riding and unjust enrichment is that the paying passengers on the bus were willing to spend the price of the ticket even though the cost could be lower had the free rider also paid for the trip. But current developing nations do not show any willingness to pay the extra bill imposed on them as a result of the behavior of the unjustly enriched countries.

Peter Birks, a contemporary legal scholar, enumerates five conditions of unjust enrichment: (1) one party is benefited; (2) this benefit is *at the expense* of the other party; (3) the enriching benefit is unjust;[12] (4) the harmed party has a right to restitution; (5) there are some defenses to which the enriched party may appeal (Birks, 2003: 6–7). Birks emphasizes that in unjust enrichment the plaintiff does not

[12] It is not strictly speaking the enrichment (in the sense of the original acquisition of the good) which is unjust, but the *retention* of the benefit or profit once the unjust nature of the holding becomes known to the parties concerned. This is highly relevant to the particular case of industrialization and its harms, which were not known either to the ancestors of the current "plaintiffs" or to the ancestors of the current "defendants."

have to prove that any "wrong" took place or that any fault could be ascribed to the defendant. The defendant, the cause of the present unjust condition, acted innocently as in the standard textbook example of paying a debt twice without any party being aware of the fact at the time. Furthermore, the unjust enrichment can be caused (as it often is) by the transfer of property, but it need not be so. Finally, Birks says that unjust enrichment should not be strictly speaking considered to be part of tort law but rather a category of its own and hence the recovery of the party sustaining loss from the one who sustained gain should be called "restitution" rather than "compensation" (the latter being reserved for the context of direct harms).

We can now see that the analysis of the ethical problem of the distribution of the burdens of global warming fits naturally into the framework of the law of unjust enrichment. Although it is hard to quantify the gains of the industrialization of the past two centuries to the Western countries, these are huge and a substantive part of these profits can be associated with pollution.[13] Thus, Birks' first condition is satisfied: the industrial countries have been enriched.[14] It should be emphasized that had we not enjoyed the fruits associated with greenhouse gas emissions, we could not be said to have been unjustly enriched and hence would owe nothing to the developing world for the so-called sins of our polluting ancestors. But whether this enrichment occurred *at the expense* of the poorer countries, as the second condition demands, is a little harder to show. After all, the industrial societies did not actually take anything (property, land, goods, labor) from the other countries. Yet, since, as we noted, unjust enrichment does not have to involve a transfer of

[13] Shue emphasizes the causal link between past industrialization and current benefits of Western countries, but does not refer to the condition of the enrichment being at the expense of the developing nations (after all, there might be a causal link between the scientific revolution and the present gap between rich and poor nations, but that does not necessarily mean that the present beneficiaries of this revolution are not entitled to the gains associated with it) (Shue, 1999: 536–537).

[14] For a somewhat similar analysis, see *Environmental Degradation, Reparations, and the Moral Significance of History* (Caney, 2006). Caney discusses the Causal Account and the Beneficiary Account of compensation for the damages of global warming. He is correct in his critique of the former (although he still sticks to its basic principle, i.e., that the actual polluter pay). But he is less convincing in the critique of the latter (the Beneficiary Account). For example, he is perplexed by the problem of the existence of many beneficiaries in the various stages of industrialization and concerned with the danger that only living beneficiaries would have to pay for the benefits of their ancestors. The problem may be less serious since the benefits of our dead ancestors were passed on to us and we have proportionately more pollution-generated wealth than they had. Thus, we are paying not only for our share of unjust enrichment but also for theirs. I am aware that this kind of description of the cumulative process of enrichment is highly schematic but it is still theoretically correct. My reply to Caney's question as to who is obliged to pay in cases when both the party causing the damage and an "innocent" beneficiary benefit from the damaging act (ibid.: 473) is that it is the person who is responsible for the damage, if only she is alive; both that person and another beneficiary in proportionate parts, if both are alive; and the "innocent" beneficiary who is alive, if the harming person is not. Note again that on my group-based account we may assume that benefits of one generation of the relevant group are passed on to the next generation. This solves the distributive problem that Caney raises regarding the alleged unfairness of beneficiaries who leave for their descendants the duty of compensation for their unjust enrichment (ibid.). Electricity (Caney's example) is not a static good that made people in the late nineteenth century better off, but a technology that gave all following generations *growing* cumulative gains that were themselves passed on to the future. I thus believe that the account in terms of unjust enrichment obviates Caney's critical arguments against the "Beneficiary Account."

property, we could easily describe the respective loss and gain in terms of the "absorptive capacity of the atmosphere." This is the public good that was used by some countries at the expense of others: "at the expense" in the sense that although at the time the good seemed to be boundless, *now* we know that much of that good has already been "consumed." The third condition is controversial and is the exact issue of contestation in the international forums dealing with climate change: was this enrichment unjust? Here we are back at the questions of group responsibility, the NIP, the problem of trans-generational relations, and the moral weight of the "bad luck" of those countries that arrived late on the scene of industrialization. But there is a compelling argument of justice that poor countries cannot be expected now to take part in "cleaning the global mess" created by others (a phrase that ironically was used by George W. Bush to justify the objection to the demand that the United States led the cleaning enterprise!).[15] The injustice of the over-use of the atmosphere consists of the *irreversible* undermining of the opportunities of later generations to use that good (which is much worse than getting hold of much of the Earth's land, since land can be redistributed at any time in a way that would be more efficient and more just). As for the fourth condition, in the context of emission history it is more accurate to speak of restitution rather than of compensation (that connotes blame for wrongful behavior). The point is that our ancestors in the West cannot be held liable for the damage of which they were not aware, and we cannot be expected to compensate for their actions. It is *we* who are "liable" because of being benefited and are called to return some of the benefit by way of restitution.[16] The "polluter pays" principle, which calls for compensation, sounds plausible but does not make sense when most of the polluters of the last two centuries are dead. The fifth condition, that of "defenses" of the beneficiaries of pollution, gives us the conceptual framework for dealing with countries such as Japan and South Korea which joined the polluting world only relatively late, or Switzerland, which has alternative clean sources of energy. They are indirect

[15] The unfairness of unjust enrichment can be described also as the profit of the enriched being higher than the market value of the relevant good. See *Unjust Enrichment* (Dagan, 1997: 17). Take Davenport's and Harris' example: if you have sold my car without my consent and received for it $10,000, I can claim the whole sum from you as an unjust enrichment even if the car is really worth only $5,000. Contrast this to the case in which the car is really worth $15,000, in which case I can claim that amount only on the basis of tort but not on that of unjust enrichment (for you have been only enriched by $10,000 although you have caused me a $15,000 damage) (Davenport and Harris, 1997: 20). In the context of climate, the gap may be seen as applying to the value at the time of the original enriching act and today's value of the relevant good (viz. the absorptive capacity of the atmosphere).

[16] See *In Defence of Historical Accountability for Greenhouse Gas Emissions* (Neumayer, 2000: 185–192). I wholeheartedly follow Neumayer's historical approach in climate ethics, particularly concerning the principle of "grandfathering." Starting from now on to distribute pollution allowances on the basis of *current* degrees of pollution of different countries is even worse than the a-historical clean-slate approach that is based on equal per capita emission rights. My only reservation concerns Neumayer's use of the term "accountability," which according to my analysis cannot be applied to the current generation in the industrial world since we cannot be strictly speaking accountable for the deeds of our forefathers.

beneficiaries of the Industrial Revolution in general but have only a small share across time of the overall quantity of emissions.

The case of affirmative action seems not to fit easily into the category of unjust enrichment, mainly because of the morally straightforward fault of the exploiters and the direct harm to the victims, both parties being at least potentially aware at the time of the wrongful practice. Nevertheless, the attempt to address the current racial relations in terms of unjust enrichment has received more scrutiny than the parallel attempt in climate change. Legal scholars have examined the alternative grounds for recompense of current descendants of slaves: compensation for past wrongs done to their ancestors, or restitution for the present unjust benefits that the descendants of the white majority enjoy. Some of these legal theorists, such as Dennis Klimchuk, argue that the latter method is superior. First, it does not require the identification of the particular wrongdoer or the particular victim. Second, unjust enrichment does not even require the proof that those who sustained loss did not consent to the practice. "The object of that claim, again, is not an event in the past, but rather a state of affairs that endures to the present," to quote Klimchuk. Third, it avoids the issue whether *in their time* slave owners should have known that the practice was wrong and whether that could serve as their "defense".[17] I find these reasons persuasive and conclude from them that in climate ethics, which lacks the alternative of a blame model, unjust enrichment is even a more plausible conceptual framework than in the case of racial discrimination, at least for anyone who insists on adopting the historical approach to climate ethics.

Finally, a similar historically sensitive model for justifying present differential contribution to the struggle against global warming is offered by Göran Duus-Otterström. According to his Inherited Debt Principle those who now hold resources whose origin can be causally traced back to the uncompensated over-use of the absorption capacity of the atmosphere "have a duty to take on extra burdens" (Duus-Otterström, 2014: 456). This is indeed a principle that is close to that of unjust enrichment since it also avoids the issue of the wrongfulness of the original actions that led to the depletion of that capacity. However, it shifts the normative focus from the *benefit* of current people (derived from the past over-use) to their *duty* (which has been inherited from the intra-generational duty of past over-users to their contemporaries). For Duus-Otterström this means that current people in the West do not have a duty *to* people in the developing countries but only a duty to make an extra effort in the independently desirable enterprise of

[17] See *Unjust Enrichment and Reparations for Slavery*, particularly p. 1274 (Klimchuck, 2004: 1257–1276). Others argue that the use of the unjust enrichment argument in the context of slavery is just "strategic" and that it is potentially demeaning to the slaves and their descendants by trivializing the horrendous wrong done to them (Sebok, 2002). Without entering into this debate I wish to note that this demeaning factor does not exist in the context of climate ethics and hence does not undermine my proposal to deal with global warming in historical terms of unjust enrichment.

reducing emissions. He admits that the principle of inherited debt cannot serve as a criterion for the actual distribution of the burden of restraining global warming, which I believe is a theoretical disadvantage in comparison to the principle of unjust enrichment.

Furthermore, there is an important difference in the normative implications of the two models: according to Duus-Otterström we have a greater remedial responsibility in the case that the over-use of our ancestors was large and the benefits to present people small than in the case that the over-use was small but its benefits large. I find this conclusion quite counter-intuitive and inevitably leading us back to some notion of an original wrongful act (which is wrong even if no one gained much from it). The unjust enrichment framework leads to the opposite conclusion: what decides the extent of our duty now is the size of our actual benefit from the original act, however (unintentionally but objectively) harmful it turned out to be. This fits well with the idea that a country that at some stage in history ("innocently") took part in polluting the atmosphere but has lost, for some reason or other, all its benefits should not be held as having a special duty today to restrain its emissions. The debt model requires that I pay for the debts of my dead father even if I am destitute or in worse shape than the creditor. In contrast, according to the unjust enrichment model I have a duty of restitution or disgorgement only if I have been *actually* enriched (at the cost of another party). And if, as Duus-Otterström seems to be saying, what incurs the *current* duty lies in the resources rather than the people having them, then the concept of "debt" is misleading since only people have debts and can inherit them. Resources passed on in history cannot be "tainted" (in Duus-Otterström's terms) independently of normative relations between actual people.

Some Difficulties

Calculating costs involved in global warming is, as everyone admits, a very complicated enterprise. Some of the problems are normative (e.g., the baseline from which distributive principles are applied). Others are empirical (what is the reasonable cap to be put on emissions or temperatures, or how much greenhouse gasses various countries emit and what part of their relative wealth or poverty is due to the scope of their emissions).[18] But it is certainly even more perplexing to try to work out differential *historical* responsibilities since it is hard to calculate the

[18] Again, the comparison to slavery is illuminating. In his detailed and nuanced analysis of the ways to measure compensation for slavery, Schedler points to the issue of the relevant baseline for the calculation of compensation. He reaches the conclusion that no individual in the present can be said to owe such compensation although the federal government may have some such responsibility (since it could have prevented the institution of slavery). But the whole analysis assumes the liability of the slave owners (and the government), which is lacking in the history of pollution (Schedler, 2002).

cumulative pollution a country has produced in the last two centuries and, again, the proportion of the current wealth of countries which is due to this pollution to their overall wealth (e.g., Switzerland vs. Britain). Furthermore, as has been emphasized by Margalioth, the contribution to the reduction in the absorptive capacity of the atmosphere must take into account not only the emission of CO_2 but equally deforestation (Margalioth, 2012: 67). How do we compare the damage of the addition of a new power station every week in China to the number of trees cut every year in Brazil? And should these computations extend to the long-distance past?

These problems are particularly disturbing if we adopt, as I suggest we do, the analysis of historical responsibility of the industrial nations through the concept of unjust enrichment. Legal scholars admit that there is something vague in the definition of "enrichment." In affirmative action this difficulty is obvious. How many minority students should be admitted every year through inverse discrimination policies in order to compensate the minority group for past injustice that benefited the majority? Similarly, how much of the enrichment of the Western countries is due to the Industrial Revolution and how much is due to colonial exploitation (or, alternatively, to scientific tradition, work ethics, or religious views)? We must also note that as long as there is no real effort of mitigation on the global level and the Chinese are continuing to emit massive amounts of greenhouse gasses, the responsibility of the historical polluters is decreasing since the new polluters are making rapid use of their opportunity right to use cheap sources of energy as Western countries have in the past. Rather than putting constant pressure on the poorer countries to slow their own emissions, the minimal kind of restitution consists of letting them make use of more polluting technologies for a limited period while restricting substantially the rates of emissions in the developed countries.

The theory of unjust enrichment always includes "defenses." In the context of pollution history these could include harms to themselves sustained by the polluters, which both in the past and in the present (and future) industrial societies suffer in terms of health risks or damage to their natural environment. And as has been repeatedly noted, there was a "spillover" of some of the benefits of industrialization to non-industrialized countries for quite a long time. These factors should offset the other benefits or profits sustained by the industrial nations.[19] (Note that these spillovers hardly exist in the parallel case of racial relations and

[19] One of the readers of this chapter suggested that the benefits provided by the industrialized world to the developing countries in terms of intellectual capital, technology, and direct aid may be so enormous so as to offset their "enrichment" by the over-use of the absorptive capacity of the atmosphere. I admit that this could be the case although most theorists, including those who oppose any restitution or compensation for past emissions, do not adhere to that opinion.

affirmative action.)[20] The defense of the northern countries that they should get a larger share of emission permits since they have higher heating costs due to cold weather can be answered by pointing to the equal costs of air conditioning in hot climates as in India and Africa. Another kind of defense is related to habituation. Industrial countries have been so accustomed to their technological achievements and material quality of life that a radical and sudden restitution would destroy their way of life and social fabric. This might imply that the reduction of emissions can be only gradual and extending over more than one generation.

A more principled version of the habituation objection is Waldron's famous idea that historical injustice is often "superseded" (Waldron, 1992: 22–23 in particular). Although the aboriginal inhabitants of Australia acquired water wells in a Lockean legitimate way, that is to say, leaving "enough and as good" to other people, at a certain stage of demographic change in the continent water became so scarce that the abundant aboriginal wells had to be shared with the new immigrants. Accordingly, the justice of acquisition has been superseded in time. The same, adds Waldron, applies to the procurement of land from the aboriginal tribes by the European settlers: it was originally unjust but with time the property has become necessary for the satisfaction of the basic needs of the descendants of the settlers, and wholesale restitution would have catastrophic consequences for them. We could use Waldron's argument for both examples of supersession of justice for climate ethics. On the one hand, even though eighteenth-century English and American industrial societies legitimately made use of the absorptive capacity of the atmosphere (leaving what at the time was "enough and as good" for others), what is left of this capacity today must be shared with others who by now have too little left to them.[21] On the other hand, the industrial societies have the analogical defense: even if the original acquisition was unjust (in reducing the pollution opportunities of others), by now the descendants have become so dependent on the acquired "good" (namely, gas emissions) that without it they will not be able to

[20] One cannot say that the black population in the United States has been indirectly benefited by slavery and discrimination. Indeed some people say that the conditions of today's African American community are superior to those of today's Africans in their countries of origin, but here the NIP arises and invalidates the comparison itself.

[21] On the Lockean way of approaching the problem of just emission rights, see *One Atmosphere* (Singer, 2010: 187). The analysis in terms of unjust enrichment assumes that our ancestors' behavior proved to be a violation of the Lockean proviso, but there is a debate whether the use of the atmosphere's absorptive capacity is analogical to the use of land, both being public goods that are not boundless. Hyams argues that industrial pollution degraded the atmosphere rather than improved it and hence, unlike in the case of land, cannot be defended by the Lockean justification of appropriation (Hyams, 2009: 247). Note, however, that the good we are speaking of here is not the atmosphere itself but the absorptive capacity of the atmosphere, and the use of *that* – it could be argued – has benefited humanity, the alternative being not exploiting *any* of this absorptive capacity, which I want to suggest would have been considered by Locke himself as pure *waste*! Luc Bovens takes a similar line by arguing that at least in the beginning of industrialization the principle of leaving "enough and as good" was not violated, and hence any attempt to appeal to it must do so across historical time (Bovens, 2014).

satisfy their basic needs. According to this argument, changes in circumstances weaken historically oriented claims and strengthen the case for a distributive, "clean-slate" approach.

Conclusion

Here is a telegraphic conclusion of the argument of this chapter:

1. Past polluters cannot be morally blamed for the current ecological crisis, but we, as beneficiaries of their actions, owe those who have been harmed *restitution*.
2. Present people, who have known for at least a few decades of the risks of continuing pollution, owe developing countries *compensation*.
3. Future people, after having settled the historical accounts of restitution and compensation, will have to share the burdens of mitigation and adaptation on the basis of *distributive justice*.

The historical approach to climate ethics cannot and must not be brushed aside. In a sense even very generous policies of redistribution of wealth and technology to the developing countries may not do full justice to the historically based global inequality that is associated with long-term pollution. Even if the historical and a-historical methods of deciding the right way to face the challenge of global warming lead to the same *actual* result in "dollar value," the *moral grounds* of the two methods are different, and overlooking the history of industrial pollution is in *itself* morally wrong.

This is not merely a philosophical issue of competing theoretical justifications but a matter of public attitude and perception. It is quite plausible that the problem of global warming will not be resolved by ethical arguments and appeals to justice, but ultimately by the universal interest of preventing a universal catastrophe and compromises based on relative power.[22] The problem of climate change is similar to the Tragedy of the Commons. Some countries are suspicious of others for not contributing their share to the preservation of the public good. Other countries have a closer access to that public good and at least in the shorter run believe that they can profit by not cooperating. But everyone knows that global warming is going to affect every society on the globe. Some argue that although the United States is currently the most significant obstacle to achieving an international agreement on emissions, it is in danger of suffering from the consequences of climate change (hurricanes, drought, flooding, extreme weather conditions) no less than poorer nations. Now my suggestion is that recognizing the historical dimension of the

[22] Traxler, who advocates the historical approach to climate ethics, is also aware that practically it would be hard to reach an agreement on the distribution of the burdens of mitigation and hence that an a-historical principle of a fair division of opportunity costs should be adopted (Traxler, 2002: 134).

current distributive problem can serve slightly to promote a cooperative, less confrontational mood. It may advance some measure of trust. As Neumayer argues: although it is true that there is no persuasive solution to the exact amount of money owed to the developing world for historical emissions, it must be acknowledged that the rich countries have to pay *something*. This is not just a symbolic acknowledgement (like apologies in cases of historical wrongs), since the historical debt is not associated with guilt. It is an objective duty that can also help in the pragmatic resolution of the climatic Tragedy of the Commons. The history of compensation in the international realm (after World War I or World War II) illustrates that although there is some measure of arbitrariness in the actual sums paid, compensation serves a symbolic purpose but also has a practical function.

Back to the ornithological simile of unjust enrichment in the book of Jeremiah:

> As the partridge hath hatched eggs which she did not lay:
> So is he that hath gathered riches, and not by right.

The second half of this verse contains what could be used as the ultimate warning to the unjustly enriched regarding climate policies: unjust enrichment backfires.

> In the midst of his days he shall leave them,
> And in his latter end he shall be a fool [villain].

References

Appiah, K. A. (2011). Group Rights and Racial Affirmative Action. *Journal of Ethics*, **15**, 265–280.

Babylonian Talmud, Tractate Baba Kama, folio 20a-b.

Birks, P. (2003). *Unjust Enrichment*. Oxford: Oxford University Press.

Bovens, L. (2014). A Lockean Defense of Grandfathering Emission Rights. In *The Ethics of Global Climate Change*, ed. D. G. Arnold. Cambridge University Press, pp. 124–144.

Broome, J. (2012). *The Public and Private Morality of Climate Change*. The Tanner Lectures on Human Values. University of Michigan, March 16. URL: http://tannerlectures.utah.edu/Broome%20Lecture.pdf.

Caney, S. (2006). Environmental Degradation, Reparations, and the Moral Significance of History. *Journal of Social Philosophy*, **37**, 464–482.

Dagan, H. (1997). *Unjust Enrichment*. Cambridge: Cambridge University Press.

Davenport, P., and Harris, C. (1997). *Unjust Enrichment*. Sydney: Federation Press.

Duus-Otterström, G. (2014). The Problem of Past Emissions and Intergenerational Debts. *Critical Review of International Social and Political Philosophy*, **17**, 448–469.

Dworkin, R. (1977). *Taking Rights Seriously*. Cambridge, MA: Harvard University Press.

Dworkin, R. (2000). *Sovereign Virtue*. Cambridge, MA: Harvard University Press.

Gardiner, S. M. (2010). Ethics and Global Climate Change. In *Climate Ethics: Essential Readings*, ed. S. Gardiner, S. Caney, D. Jamieson, and H. Shue. Oxford: Oxford University Press, pp. 3–38.

Gosseries, A. (2004). Historical Emissions and Free-Riding. *Ethical Perspectives*, **11**(1), 36–60.

Hyams, K. (2009). A Just Response to Climate Change: Personal Carbon Allowances and the Normal-Functioning Approach. *Journal of Social Philosophy*, **40**(2), 237–256.

Heyd, D. (2009). The Intractability of the Nonidentity Problem. In *Harming Future Persons*, ed. M. Roberts and D. Wasserman. Dordrecht: Springer, pp. 3–25.

Klimchuk, D. (2004). Unjust Enrichment and Reparations for Slavery. *Boston University Law Review*, **84**, 1257–1276.

Margalioth, Y. (2012). Assessing Moral Claims in International Climate Change Negotiations. *Journal of Energy, Climate, and the Environment*, **3**, 42–80.

Margalioth, Y. (2013). Analysis of the US Case in Climate Change Negotiations. *Chicago Journal of International Law*, **13**(2), 489–505.

Meyer, L. H., and Roser, D. (2006). Distributive Justice and Climate Change: The Allocation of Emission Rights. *Analyse & Kritik*, **28**, 223–249.

Miller, D. (2008). *Global Justice and Climate Change: How Should Responsibilities be Distributed?* The Tanner Lectures on Human Values. Beijing: Tsinghua University, March 24–25. URL: http://tannerlectures.utah.edu/_documents/a-to-z/m/Miller_08.pdf.

Mora, C. et al.(2013). The Projected Timing of Climate Departure from Recent Variability. *Nature*, **502**, 183–187.

Neumayer, E. (2000) In Defence of Historical Accountability for Greenhouse Gas Emissions. *Ecological Economics*, **33**, 185–192.

Nozick, R. (1974). *Anarchy, State, and Utopia*. Oxford: Blackwell.

Page, E. A. (2012). Give It Up for Climate Change: A Defence of the Beneciary Pays Principle. *International Theory*, **4**, 300–330.

Posner, E. A. and Weisbach, D. (2010). *Climate Change Justice*. Princeton, NJ: Princeton University Press.

Rawls, J. (1999). *The Law of Peoples*. Cambridge, MA: Harvard University Press.

Schedler, G. (2002). Principles for Measuring the Damages of American Slavery. *Public Affairs Quarterly*, **16**, 377–404.

Sebok, A. J. (2002). *The Brooklyn Slavery Class Action: More than Just a Political Gambit*. April 9. URL: http://writ.news.findlaw.com/sebok/20020409.html.

Shue, H. (1999). Global Environment and International Inequality. *International Affairs*, **75**, 531–545.

Singer, P. (2010). One Atmosphere. In *Climate Ethics: Essential Readings*, ed. S. Gardiner, S. Caney, D. Jamieson, and H. Shue. Oxford University Press, pp. 181–199.

Traxler, M. (2002). Fair Chore Division for Climate Change. *Social Theory and Practice*, **28**(1), 101–134.

Waldron, J. (1992). Superseding Historical Justice. *Ethics*, **103**, 4–28.

2

Historical Responsibility and Climate Change

JANNA THOMPSON

One of the sticking points in negotiations between developed and developing countries about the contributions they should make to alleviate the effects of climate change is whether developed countries should bear the greater share of the burden because of their historical responsibility for causing the problem.[1] Since the Industrial Revolution the activities of people in these countries have been contributing to the buildup of greenhouse gases (GHGs) in the atmosphere. It is this historical accumulation of harmful emissions that has led the world to the crisis that it is now facing. Henry Shue compares our present situation to one in which some individuals hog all the places in a parking lot.

> The billion or so poorest human beings on the planet need sound and sustainable economic development. They need "space" for their increased emissions to the earth's atmosphere ... They need to use the emission absorptive capacity, but no absorptive capacity is left because those of us in affluent economies have taken it all (and much more). We are parked in their spaces and no empty spaces exist.
>
> *(Shue, 2001: 452)*

There are two injustices that Shue points out by the use of this analogy and each has a double aspect. The first focuses on present relationships. People (mainly but not only) in developed countries are using more than their fair share of something that everyone needs. In doing so they are not only unfairly monopolizing a global resource. The harms caused by GHG emissions will be visited primarily on those who did little or nothing to contribute to the problem. The poor of the world are particularly vulnerable to rising sea levels, extreme weather, water shortages, and other consequences of climate change. The other injustice is a historical wrong. The absorptive capacity has been used up and harmful levels of GHGs have

[1] Brazil proposed to the United Nations Framework Convention on Climate Change (1997) that countries that have contributed the most to climate change bear the greatest responsibility in combating it. See La Rovere et al., 2002.

accumulated in the atmosphere because of what people in developed countries did in the past. This injustice would remain even if developed countries were now prepared to cut their emissions. It would remain even if they were prepared to compensate people in poor countries for the harm their own emissions are causing.

It is a deeply entrenched moral conviction that those who are morally responsible for unjustified harm have an obligation to make reparation to their victims. The polluters should pay. Nevertheless, demands for reparative justice for historical emissions run into serious problems concerning agency and moral responsibility. Two seem particularly damaging to these demands.[2] The first is the problem of ignorance. Agents causally responsible for emissions in the nineteenth century and most of the twentieth century did not know that they were using up a global resource or that they were contributing to a problem that would have severe effects on people in other countries (as well as on future generations). They were not guilty of deliberately causing these harms. Nor can they be accused of recklessness or negligence. These failings presuppose either that agents knew of the risks associated with their behavior or that they were in the position to acquire this knowledge. Most of those concerned with climate justice conclude that responsibility for climate change did not exist before people in developed countries became aware of the problem: in the late 1980s or in 1990, when the first report of the Intergovernmental Panel on Climate Change was released (Caney, 2010 and Baatz, 2013).

The second problem relates to the assignment of reparative responsibility. Existing people in developed countries are responsible for the harm caused by their own emissions, but they bear no causal responsibility for the emissions of their predecessors. They were not the historical polluters. How then can we justify assigning them reparative responsibility for what past people and governments did (Baatz, 2013: 95–96)?

These objections depend on two basic assumptions about justice and responsibility: that responsibility for reparation belongs only to the agents who did the injustice, and that an act is unjust only if the agent is culpable – that is, only if she knew or ought to have known that she was acting unjustly. Agents who are excusably ignorant about the harm they are causing cannot be culpable. And they cannot be liable for reparation if they were not the ones who caused the harm. From these assumptions it follows that present agents, whether individuals or governments, cannot owe reparation for harms caused by GHG emissions of

[2] These objections, and others, to the "polluter pays principle" are presented by Caney, 2006 and 2010; Page, 2008; Baatz, 2013; and many others. They constitute two of the three objections to the claim that historical emissions should play a role in determining responsibility for climate change referred to in section 3.3.4. of the Contribution of Working Group III to the Fifth Assessment Report of the Intergovernmental Panel on Climate Change (Kolstad et al., 2014). The third, the problem of nonidentity, is discussed very briefly later.

historical agents, first of all, because they or their present members played no role in their causation and, second, because past agents, being excusably ignorant, did not do an injustice for which reparation could be owed.

These considerations have caused most philosophers to doubt that reparative justice applies to the emissions of past generations. Some justify demands on the rich by citing their ability to pay or by appealing to requirements of distributive justice. Lukas H. Meyer and Dominic Roser conclude that the duty of the rich of the North to the poor of the South should be conceived mainly as an obligation of distributive justice (Meyer and Roser, 2010: 247). Caney argues that historical emissions are significant only because their existence should provide a motivation for sharing resources more equally (Caney, 2006: 477). Others argue that the case for compensation for historical emissions rests not on the harms done by emitters but on the benefits that present people have received from these emissions. Alex Gosseries argues that by receiving benefits from activities that harm others, people in developed countries are free riding (Gosseries, 2004). They are not paying their share of the costs and should be prepared to surrender the benefits they have received from being free riders. Caney argues that those who receive benefits from historical emissions are liable for compensation (Caney, 2010).

The idea that those who benefit from an injustice or an unfair relationship ought to compensate those who are harmed is widely accepted. But it is difficult to justify demands for compensation from those who did not receive benefits voluntarily and (in many cases) had no real chance to refuse them (Meyer and Roser, 2010: 235). The accusation of free riding is most applicable in cases when participants are deliberately taking advantage of a scheme from which they derive benefits or are, at least, aware that there is a scheme from which they are benefiting. Those who abandon appeals to historical justice in favor of the demands of global distributive justice are relying on an idea that has limited appeal outside the circles of those committed to moral cosmopolitanism. As Caney remarks, an approach that is concerned only with achieving equity in the distribution of resources runs contrary to a widely shared assumption that the historical origin of a problem must play an important role in the assignment of responsibility. "A wholly forward-looking approach, it might be argued, is out of kilter with some of our deepest moral convictions" (Caney, 2010: 214).

Those concerned with climate justice have reason to investigate all available approaches. But to abandon the appeal to reparative justice for historical emissions is to give up on an intuitively powerful idea of why people in rich countries owe compensation to the poor. If it can be defended, there is reason for doing so. In this chapter I will offer a defense of reparative justice for historical emissions by arguing that excusably ignorant agents can do injustice and thus be liable for reparations; and that present people ought to accept a reparative responsibility for the injustices of their predecessors. My defense does not attempt to overcome all of the problems of applying reparative justice to harms caused by historical emissions. Its aim is to show that the most severe objections – those that depend on widely held views about culpability and responsibility – can be defeated.

Responsibility and Culpability

That nonculpable agents should accept some reparative responsibility for causing harm to others is, according to some philosophers, a matter of common sense. Shue thinks it obvious that people should take responsibility for the messes that they make – whether they are culpable or not. "All over the world parents teach their children to clean up their own mess" (Shue, 1999: 533). This requirement, he thinks, is demanded by considerations of fairness and equity. "If there were an inequality between two groups of people such that members of the first group could create problems and then expect members of the second group to deal with the problems, that inequality would be incompatible with equal respect and equal dignity. For the members of the second group would in fact be functioning as servants for the first group" (ibid.: 536).

However, Caney argues that a requirement of strict liability assigns an unjustified priority to the interests of right holders over the interests of duty bearers (Caney, 2010: 209). The master–servant relationship, in his view, is reversed. A nonculpable person is subjected to the claim of another just because he caused an accident through no fault of his own. The same criticism can be made of David Miller's attempt to extend the scope of moral responsibility by arguing that agency can be sufficient to give a person a rectificatory responsibility for the harmful outcomes of his act (Miller, 2007: 93). If a person has taken all reasonable precautions to ensure that her bonfire will not get out of control, he says, she can nevertheless be held responsible for the damage done to her neighbor's property when an unanticipated wind causes sparks to fly over her fence. Her responsibility for rectification exists, according to Miller, because she was the agent in charge of the fire that caused the harm. But even if we agree that the victim deserves redress, it seems unfair that the unlucky agent should assume the entire burden of providing it. Another way of providing compensation for harm caused by such accidents seems preferable – perhaps through a fire insurance scheme to which everyone in the neighborhood contributes. Or perhaps the people who enjoyed the bonfire should pay. There is no moral reason why the nonculpable agent has to bear the burden for compensating the victim, and her nonculpability is a reason for not making this requirement. If no other way of providing compensation is available, then perhaps it is justified to assign the responsibility to her. But if outcome responsibility exists only to ensure that the victim receives compensation, then it is best understood, as Caney suggests, as a morally questionable convention of "strict liability" rather than the consequence of bearing moral responsible for a wrong. To mount a more effective challenge to views that tie rectificatory responsibility to culpability it is necessary not only to point out that nonculpable agents can do acts that violate others' rights but also to explain how doing so can make them morally responsible and, for this reason, liable for redress.

Shue and Miller focus on cases when nonculpable agents cause harm because they are not in control of all of the causal factors that determine the outcome of their actions. What makes it difficult to assign them responsibility is that their ignorance of these factors does not reflect negatively on their career as moral agents. Using Mathias Risse's distinction between a "time-relative" and a "time-neutral" basis for moral judgments, Derek Bell provides an epistemological explanation for why nonculpable agents should take responsibility for the wrongs that they do (Risse, 2008 and Bell, 2011). A time-relative basis for judgment is available to agents at the time of their action. A time-neutral basis is what would be available to agents if they were not limited by lack of relevant knowledge. Agents have no choice but to act according to the best reasons that their temporal position makes available. But from a time-neutral standpoint their actions may nevertheless be unjust. The Athenians thought that there was nothing wrong with slavery. Their belief was backed up by their views about human differences and capacities – indeed, by the theories of no less a philosopher than Aristotle.[3] But we have good reason to believe that slavery is and was wrong and that the biological theories that supported it were mistaken. The Athenians were not culpable for holding slaves. They were acting justly according to their time-relative perspective. They lacked the empirical and moral knowledge that would have made them able to judge slavery to be wrong. Nevertheless they acted wrongly (Bell, 2011: 405).

Bell applies his account of moral responsibility to historical emissions. People in the nineteenth and early twentieth centuries had no way of knowing that their emissions were contributing to harmful climate change. Their time-relative basis for judgment did not provide them with this information. They did not know that they were using up a resource that others would depend on or were contributing to a state of affairs that would cause harm to others. But from a time-neutral perspective they were committing an injustice and this is sufficient to give them a reparative responsibility.

To accept Bell's justification for retrospective responsibility we need to understand why a person who acts conscientiously from a time-relative perspective ought to accept moral responsibility for what is unjust on a time-neutral basis.

[3] Aristotle argued in the first book of his *Politics* that some people are slaves by nature.

Then we need to determine whether these reasons apply to the case of climate change where lack of knowledge is empirical rather than moral.

Suppose a slave owner acquires the moral and empirical knowledge that he lacked and now believes that slave owning is wrong. Given that his former ignorance was excusable, why should he now accept responsibility for wrong-doing? Bell appeals to the motivations of moral agents. Conscientious agents want to do what is right. But they know that their knowledge is imperfect and that they may be acting wrongly. If advances in knowledge reveal to them that they did wrong, then they will want to put things right. The obligation of a blameless agent to make redress is not as great as it would be if she were culpable. But at least she should not want to retain the benefits of her unjust acts.

> She cannot change the past. However, if she sincerely regrets that she has acted wrongly, she should not want to have benefited from her wrongful act. Therefore, she should be willing to accept that she should not retain the benefits derived from her wrongful acts. Instead, these benefits may be transferred to the victims of her wrongful acts to rectify (or partially rectify) the wrong that she has done.
>
> *(Bell, 2011: 403)*

Bell's explanation of how a nonculpable agent can be morally responsible for reparation gains plausibility from the fact that the source of the wrongdoing is not so much in external factors but in the agent himself. An epistemological lack is a failing that belongs to the agent even if he is not responsible for it and thus there is better reason to assume that responsibility for repair belongs to him. This reasoning is bolstered by a consideration of what it means to be a self who responds appropriately to past deeds that affect her conception of who she is as a moral person. A responsible moral agent is someone who is able and willing to take responsibility not just for present actions but also for the actions that belong to her history as an agent. Being someone who can exercise this responsibility is crucial to maintaining proper relations with others. It enables her to be worthy of trust – to be regarded by others as someone who treats being just as a primary objective of her moral life. It entitles her to think of her life and deeds as contributing to a state of affairs in which justice is done and injustices are acknowledged and repaired. It is also important for her conception of who she is. By acknowledging and accepting a duty of repair for injustices that result from her failings as an agent, including nonculpable failings, she dissociates herself from the injustice. That she did what she did is undeniable, but by acts of repair she changes her relation to her past actions and thus retains respect for herself as a moral individual for whom just action is a guiding ideal.

This conception of the moral individual supports Bell's insistence that a person who finds that she has committed an injustice will not want to retain the benefits of her deed. But an emphasis on the moral taint that results from past misdeeds cannot provide the full story. If the wrongdoer is motivated to divest herself of the benefits merely because she thinks they are tainted by the wrong, then there is no particular reason why she should transfer them to those who were harmed. She might decide that she ought to give them to the least well off. An adequate explanation of why a nonculpable agent can be morally responsible for reparation to those harmed also requires an account of the effect of wrongdoing on moral relationships.

An injustice is an injury that undermines relations that ought to exist between people. Intrinsic to the failure to respond appropriately to an injustice is a lack of proper respect for others. An agent who does not respond appropriately has, from the victims' point of view, treated them as if the harms the injustice caused do not matter. Repair of an injustice thus calls for an appropriate response from the doers of wrong – an acknowledgment that their behavior was unjust and acts of reparation that aim to repair relationships damaged by the injustice. Since good moral relations between agents depend on a willingness to act justly and to acknowledge and repair injustices, agents who know that they can commit, or suffer, injustice because of the time-relative ignorance of them or others have reason to accept a moral requirement to make reparation to victims of unjust acts – a requirement that does not depend on whether they knew that they were acting unjustly. We would expect the former slave owner to recognize this requirement.

These considerations provide an answer to Caney's contention that demanding reparation from nonculpable agents is to give undue precedence to rights holders. An injustice not only violates the rights of others. It also undermines moral relationships. Reparation is not a matter of merely compensating victims but also of repairing these relationships. Even nonculpable agents can have this responsibility.

However, this conclusion seems to depend on the nature and extent of the ignorance that stands in the way of right judgment. Suppose that what is right from a time-neutral perspective is very different from what we now believe is right and we have no way of finding this out. We are utterly mistaken in our moral judgments, and our ignorance, at least for the time being, is irreparable. If our ignorance is so deep and complete, then it is not plausible to suppose that we have any moral responsibility for the wrongs we do or that others have any reparative claim on us. After all, if ignorance is total and we cannot overcome it, then how can our lack of knowledge reflect on us as moral persons or undermine our relations with others?

However, it is unlikely, even incomprehensible, that the time-neutral perspective could be so different from our time-relative perspective. We know that morality requires respect for the feelings, interests, and points of view of others; we cannot imagine or accept a time-neutral view that tells us something different.

If we are mistaken in our moral beliefs, then most likely this is because we have not paid sufficient attention to the interests or views of others or because we fail to search for evidence that might contradict our empirical assumptions. An ideal moral agent would not have these failings. We are not ideal moral agents and thus our ignorance is often excusable. Nevertheless, our failure to comprehend or look for information that could reveal our ignorance is a moral failing even if it does not make us culpable, as this is usually understood. The existence of this failing, I contend, does the work of explaining why an agent can sometimes be held morally responsible for repairing wrongs that she did out of excusable ignorance.

However, this way of understanding how moral responsibility can be attributed to ignorant agents raises doubts about Bell's attempt to apply his account of responsibility to climate change. The lack of knowledge of people of the nineteenth and early twentieth centuries about the effect of GHG emissions on climate was total. They had no idea that their emissions could cause this problem. Scientific data or theories that could have alerted them were simply not available. The time-neutral perspective was well beyond the scope of their science. But if they were completely ignorant, then it is difficult to understand how they can be held morally responsible for their contribution to climate change.

Nevertheless, moral and epistemological ignorance often go together. False theories, lack of relevant data, or misinformation often underwrite moral injustices – as in the case of slavery. And false moral views or moral failings can also contribute to epistemological failure by discouraging agents from seeking out new information or questioning existing theories or by encouraging them to engage in risky behavior. Alexa Zellentin argues that people who engaged in emission generating activities in the nineteenth and early twentieth centuries should have known that they were doing something that involved unforeseen risks (Zellentin, 2015). In her view, people who engage in activities when lack of knowledge makes the consequences of their actions unforeseeable have a duty of care to those who might be harmed, and thus a responsibility to make reparation if harm occurs, even if they were justified in acting as they did. People in the decades after the Industrial Revolution were ignorant of the contribution of their emissions to climate change. But, says Zellentin, "they should have recognised that they could not yet foresee what might follow from such unprecedented interference with the climatic system" (Zellentin, 2015: 264). The fact that they acted outside their competence to foresee and control outcomes means that they owe reparation for their contribution to the harm that eventuated.

Zellentin depends on a distinction between actions carried on within our sphere of competence, where we have good reason to believe that we understand the risks and can control outcomes, and actions outside our sphere of competence – unprecedented activities that could result in unforeseeable harm to others. If we perform the former conscientiously but nevertheless cause harm (as in the case of the bonfire), then we are merely unlucky and this is what makes demands for reparation seem unfair. Undertaking the latter can make us responsible for reparation for resulting harm because we know that we are not competent to control outcomes but we act anyway. The problem for her account of responsibility is that any activity that is not simply a repetition of what has been done before is unprecedented and carries unknown risks. Indeed, all our acts have unprecedented aspects. The bonfire lighter may be very competent with bonfires, but she has never lit one at this time and place and has to admit that there could be unknown risks. Zellentin's account is unhelpful if it does not explain why some unprecedented acts make us morally responsible for harm and others do not.

Supplemented by additional considerations (some of which Zellentin provides herself), her account nevertheless makes a good case for insisting that our nineteenth- and early twentieth-century predecessors, despite their ignorance, bore reparative responsibility for their contribution to climate change. Everyone at that time knew that emissions of various kinds were an inevitable consequence of the means of production. Almost everyone was aware that these emissions caused harm. They caused lung diseases in industrial workers. They created health problems in industrial cities. They had obvious destructive effects on the environment, and it was reasonable to believe that emissions were also causing other, less visible harms that had not been monitored or investigated. Our predecessors knew nothing about the effect of GHGs on the Earth's atmosphere. But their knowledge of the harms emissions did cause and their awareness that they were likely to be the cause of as-yet-unknown harms, gave them good reason to fear that their activities would result in unforeseen harm to others. They should have been more conscientious in investigating the effects of their emissions. If they had, then a science able to recognize the effect of GHG emissions on the climate may have come into existence earlier. They should have taken the problems known to be caused by emissions more seriously (and not regarded them as an unfortunate but acceptable price of progress). If they had, then they would probably have done more to limit emissions and find alternative ways of producing goods. Their ignorance of the effect of GHGs on the climate is excusable. Their failure to take the problem of emissions more seriously is also probably excusable given their cultural beliefs and the economic forces that acted on them. Like the slave owners of ancient Athens, they were not ideal ethical agents and could not reasonably have been expected to think and act as ideal agents. Nevertheless, their moral failures and their responsibility to those they put at risk of harm by engaging in activities that they knew were productive of harm are sufficient to give them reparative responsibilities – not only to people they knowingly harmed by their emissions but also to those harmed by the unforeseen effects of their emitting behavior. As in the case of the former slave owner, their understanding of themselves as moral agents and their moral relations to others required them to accept this responsibility.

Responsibility for Reparation

However, the fact that the emissions are historical – the result of activities of people of past generations – gives rise to the second major objection to demands for reparative justice for historical emissions: that it is unjust to demand reparation from people who were not even in existence when the emissions occurred and thus had no role in causing them. Those like Shue who think that reparative demands for climate change are justified generally assume that the agents liable for reparation are rich nation-states and that they owe reparation to poor nation-states. States are associations organized in a way that makes them capable of acting as agents, and they persist through a number of generations. So assigning to them responsibility for reparation is an obvious way of meeting the objection. Giving them responsibilities to poor nation-states rather than to harmed individuals also answers another common objection to demands for reparation for historical injustices: that individuals whose existence depends on the unjust activity cannot be said to be harmed by it since no comparison can be made between their present situation and how they would have fared if the injustice had not been done. Moreover, regarding nation-states as the relevant agents seems more appropriate than attempting to assign responsibilities to individuals or other collectives. Individuals and even corporations often had little choice about their participation in emitting behavior. States in wealthy countries provided the framework of laws and conventions that encouraged the development of industrial production and tolerated lack of attention to external costs. Industrial states were in the best position to gather information about these costs and to do something about their causes. This does not mean that other agents bore no responsibility. It does not mean that rich polluters in poor countries cannot owe reparations. But it justifies an emphasis on the responsibility of nation-states for historical emissions.

However, assigning reparative responsibility to existing rich states for historical emissions means assigning responsibility to present governments and citizens. They are the ones who have to pay the price for what their predecessors did or failed to do. But since they had nothing to do with causing historical emissions it seems that they have a legitimate objection to this burden. A refusal to accept it is backed up by a common idea about collective responsibility. If participating in making political decisions or being able to participate is a prerequisite for sharing responsibility for what a state does, then present citizens share no responsibility for the injustices allowed or encouraged by their predecessors.

Present citizens did not cause the historical injustice. But it does not follow that they cannot be held responsible for reparation. Meyer and Roser provide two reasons why citizens should accept reparative responsibility in this case as well as in other cases of historical injustice (Meyer and Roser, 2010: 244–245). One of them is a concern for natural justice. A state is intrinsically valuable only if it is just. Since it is an intergenerational society, it is just only if its citizens are prepared to compensate victims for injustices, including the injustices done before their existence. There is no reason why these cannot include injustices that their predecessors, given their time-relative perspective, did not know were unjust.

> We all have the duty to support the realization of justice and thus, if membership in a community, and the creation of adequate institutions in this community, and the support of the specific goal of compensation payments to victims of historical injustice are necessary prerequisites for justice becoming a reality, then one has a duty to work towards these goals *(ibid.: 245)*

The second reason is that those who value their national community and identify with it will accept requirements of reparation for the sake of their own self-respect as citizens and for the sake of their state's integrity and respectful relationship to other states. These reasons for accepting the requirements are much the same as those that support Bell's account of why agents should accept a duty of reparation for acts for which they were not culpable.

But Meyer and Roser do not make it clear why the existence of an intergenerational state gives citizens a duty of justice to compensate for wrongs of past generations. People of wealthy countries might be under a natural duty of justice – or a humanitarian duty – to give aid to those who have been harmed by a historical injustice, but this is not the same as having a duty of reparation. Meyer's and Roser's reasoning depends on the state's being an agent whose responsibilities in respect to its historical past are similar to the responsibilities of a human agent in respect to his or her past deeds. If this analogy holds, then all generations have to accept responsibility for historical injustices of their predecessors just as individuals have to accept responsibility for their past behavior. But why should individuals accept responsibility for the deeds of predecessors? Why should natural justice provide more support for the proposition that those harmed by injustice should be compensated than to the proposition that people should not be expected to pay for what others did? Why should justice in an intergenerational society be anything more than a requirement that each generation of citizens should ensure that its own policies and practices are just and that its own injustices are repaired?

The explanation, in my view, depends on the reasonable assumption that citizens want, or should want, their state to continue to be just in the future beyond their lifetimes – for the sake of their children and grandchildren or the groups with whom they identify, or because they value maintaining respectful relations with other intergenerational communities, or simply because they value the perpetuation through the generations of just institutions and practices. If citizens have this moral concern, they will demand that their successors accept an obligation to continue to maintain and reform just institutions, maintaining respectful relations with other intergenerational communities and repairing injustices, including the wrongs of past generations. If they believe that their successors have this obligation, they must accept the same duties in respect to the deeds of *their* predecessors.[4] They have good reason not to exclude injustices done out of ignorance, for citizens of each generation know that their moral judgments are time-relative and that they can unknowingly do what future generations will, with good reason, regard as unjust. Their concern that their state should continue to be a just agent must include a concern that such injustices be corrected and victims compensated.

Having this responsibility does not depend on whether citizens identify with their state as an intergenerational community. They can have other reasons for doing their share to ensure that the state acts justly through the generations. Having a responsibility for reparation does not require that their country was completely democratic when it committed the injustice or even that it was democratic at all.[5] Many of the injustices done by a state are rooted in the culture and everyday practices of its citizens (as were, and are, the activities causing climate change). Governments, even authoritarian ones, are supposed to make decisions that reflect the values of the national culture and the interests of citizens. There are undoubtedly limits to intergenerational responsibility and cases when doubts arise about its existence. But these doubts are not likely to undermine demands for reparation for historical emissions from wealthy states.

What Is Owed?

By allowing, and indeed encouraging, the emission of GHGs into the atmosphere wealthy states were committing an injustice for which their present citizens owe reparation. They owe reparation because the injustice was done and the harm was caused and not because they gained benefits from the emitting activities of their predecessors (which does not mean that making demands of benefiters cannot be justified for other reasons). What then do wealthy states owe in reparation? The difficulty of answering this question gives rise to further complaints about reparative justice as a response to the harms caused by climate change.

[4] I defend this position in Thompson, 2002.

[5] This is one of the reasons that Caney gives for thinking that citizens of today or citizens of states that were authoritarian or only partially democratic cannot owe reparations (Caney, 2006: 470–471).

Reparative justice requires, according to standard accounts, that harms caused by an injustice be eliminated so far as possible.[6] What was taken from the victims should be returned along with compensation for any collateral losses that they suffered. The victims should be restored to the situation they were in before the injustice took place, or, if this is impossible, they should be fully compensated for the disadvantages they are suffering because of the injustice. In the case of historical emissions, these requirements pose insurmountable practical and moral problems. How much harm climate change will cause is uncertain (Caney, 2010: 206). The vulnerability of people in poor countries to its effects may have many causes – including past policies of their governments. There is no way of knowing what the situation of poor countries would be if wealthy countries much earlier in their history had made a strong effort to limit emissions. Requiring wealthy countries to provide full compensation for all of the harms caused by their contributions to climate change (to the extent that this can be determined) is not only an impossible demand to fulfil but also one that is likely to have severe, and morally unacceptable, implications for the well-being of present and future citizens of these countries. Moreover, making developed countries pay ignores the fact that some countries that did not contribute much to historical emissions are now rapidly increasing their output of GHG. Jeremy Waldron in his discussion of difficulties of applying requirements of reparative justice to historical wrongs concludes that historical injustices are superseded by time and change and that we ought to concentrate on bringing about a just distribution of wealth and resources in the present (Waldron, 1992). The problem of determining responsibilities for repair of harms caused by historical emissions makes his position attractive.

However, this problem does not mean that demands for historical justice should be abandoned. What they establish is that we need a better way of understanding what reparation requires. Injustice undermines relations of respect and trust that ought to exist between agents. I have argued that this is a reason for the existence of an obligation of repair – even for injustices that were done in ignorance. The appropriate form of reparation is, therefore, what is required to repair relationships that were damaged by the injustice or to establish relations in the present that all parties can regard as just. But this means that appropriate reparation must take into account present circumstances. Relations would not be repaired or made more just if wealthy nations had to impoverish themselves in order to pay their debt. But they would also not be repaired or made just if wealthy nations confined themselves to acknowledging a historical responsibility and did nothing to address the harms that are being caused. Just reparation has to

[6] This conception extends back at least as far as Aristotle, who in the *Nicomachean Ethics*, book 4, sec. 4 treats an injustice as an upset to the status quo that reparative justice puts right by restoring perpetrator and victim to their original position.

take into account the present interests and needs of agents. It should also take into account the fact that people in developed countries were for a long time ignorant of the effects on the climate of emissions. Ignorance is no reason for refusing to make reparation, but it may be a reason for reducing demands (as Bell believes). In general terms the parties should aim for a settlement that all have reason to agree is just, taking into account recent developments, present needs and capacities, and other demands of justice, including those that can be made of wealthy nations that are presently contributing to the problem but are not responsible for historical emissions. There is no formula for determining when justice is achieved and thus no easily determined answer to the question of what wealthy countries owe in reparation for historical emissions.

What seems reasonable for wealthy countries to provide? The answer is likely to be similar to the answer given by those who take different approaches to the question of what climate justice requires. Wealthy countries should help poor countries adapt to the unavoidable effects of climate change, and they should compensate them for its effects; they should, for example, provide assistance and resources to those who are displaced. These measures are also advocated as answering to the requirements of distributive justice, human rights, or humanitarian obligations (Page, 2013). This convergence raises the question of whether appeals to reparative justice have a distinctive or useful role in moral debates about what climate justice requires.

The existence of more than one way of defending the claim that developed countries have responsibilities to people in poor countries is no reason for thinking that all but one must be superfluous. Many philosophers have concluded that more than one approach is necessary to provide a complete account of why the rich have moral obligations to the poor. Page argues that there is no single principle or approach that identifies developed countries and their citizens, cleanly and persuasively, as the entities that should shoulder the burden of mitigation and adaptation (Page, 2011: 425ff). My defense of the existence of reparative obligations for historical emissions aims to show that the polluter pays principle can have a larger role than many of its critics suppose in identifying those who have responsibilities. It can justify reparation not only for injustices presently or recently committed but also for injustices done in earlier generations. This could have an important implication for attempts to present a moral case to citizens of wealthy countries for helping those who will be severely affected by climate change. Demands of reparative justice have a force that is hard for morally conscientious people to resist. Most can be persuaded that they have a duty to make reparation for their injustices, including the injustices of their nation-state. My aim is to provide philosophical backing for the intuitive conviction that the history of an injustice matters.

References

Aristotle (1980). *The Nicomachean Ethics*. Oxford: Oxford University Press.

Baatz, C. (2013). Responsibility for the Past? Some Thoughts on Compensating Those Vulnerable to Climate Change in Developing Countries. *Ethics, Policy & Environment*, **16**(1), 94–110.

Bell, D. (2011). Global Climate Justice, Historic Emissions, and Excusable Ignorance. *The Monist*, **94**(3), 391–411.

Caney, S. (2010). Climate Change and the Duties of the Disadvantaged. *Critical Review of International Social and Political Philosophy*, **13**(1), 203–28.

Caney, S. (2006). Environmental Degradation, Reparations, and the Moral Significance of History. *Journal of Social Philosophy*, **37**(3), 464–82.

Gosseries, A. (2004). Historical Emissions and Free-Riding. *Ethical Perspectives*, **11**(1), 36–60.

Kolstad C., Urama, K., Broome, J., Bruvoll, A., Cariño Olvera, M., Fullerton, D., Gollier, C., Hanemann, W. M., Hassan, R., Jotzo, F., Khan, R. M., Meyer, L. H., and Mundaca, L. (2014). Social, Economic and Ethical Concepts and Methods. In *Climate Change 2014: Mitigation of Climate Change: Contribution of Working Group III to the Fifth Assessment Report of the Intergovernmental Panel on Climate Change*. Cambridge: Cambridge University Press.

La Rovere, E. L., et al. (2002). The Brazilian Proposal on Relative Responsibility for Global Warming. In *Building on the Kyoto Protocol: Options for Protecting the Climate*, ed. K. Baumert, O. Blanchard, S. Llosa, and J. Perkaus. Washington, DC: World Resources Institute, 157–73.

Meyer, L. H., and Roser, D. (2010). Climate Justice and Historical Emissions. *Critical Review of International Social and Political Philosophy*, **13**(1), 229–53.

Miller, D. (2007). *National Responsibility and Global Justice*. Oxford: Oxford University Press.

Page, E. A. (2008). Distributing the Burdens of Climate Change. *Environmental Politics*, **17**(4), 556–75.

Page, E. A. (2011). Climatic Justice and the Fair Distribution of Atmospheric Burdens. *The Monist* 94 (3) 412–32.

Risse, M. (2008). Who Should Shoulder the Burden? Global Climate Change and the Common Ownership of the Earth. *Harvard Kennedy School Faculty Research Working Papers Series*. RWP08-075.

Shue, H. (1999). Global Environment and International Inequality. *International Affairs*, **75**(3), 531–45.

Shue, H. (2001). Climate. In *A Companion to Environmental Philosophy*, ed. D. Jamieson. Oxford: Blackwell, 449–59.

Thompson, J. (2002). *Taking Responsibility for the Past: Reparation and Historical Injustice*. Cambridge: Polity.

Waldron, J. (1992). Superseding Historical Injustice. *Ethics*, **103**, 4–28.

Zellentin, A. (2015). Compensation For Historical Emissions and Excusable Ignorance. *Journal of Applied Philosophy*, **32**(3), 258–74.

3

Historical Emissions

Does Ignorance Matter?

DANIEL BUTT

This chapter addresses a specific issue in relation to the question of who should pay the costs of mitigation, adaption, and compensation stemming from anthropogenic climate change.[1] It is concerned with the claim that it is in some way inappropriate to make members of current-day states pay for the historic greenhouse gas emissions (GHGs) of their counterparts because those historic counterparts were not aware of the harm to which their actions would give rise in the future. The analysis is specifically concerned with a particular form of this argument, which maintains that this historical ignorance is the key factor that makes a difference to what is owed in the present. The argument is not always made in this fashion: sometimes it is listed along with a number of other concerns relating to the practical and theoretical difficulties associated with holding present-day communities responsible for the actions of their forebears in order to cast general doubt on backward-looking principles for the allocation of present-day costs. My concern is with the particular claim that the fact of historical ignorance is key: that it exculpates historical actors, meaning that we do not, in retrospect, believe that they were to blame for their actions, and that this means that it would be wrong to hold their successors responsible for rectifying the costs of the lasting effects of what they did (cf. Bell, 2011; Zellentin, 2014). My aim in this chapter is provisional: it is to show that this specific argument does not apply to the particular context of historic emissions, and that this "exculpatory block" does not lessen or remove modern-day remedial obligations, given a belief that such remedial obligations would in fact exist had past generations been aware of the likely effects of their actions. I accept that ignorance is a defense against the charge of moral wrongdoing, absent a justified charge of culpable negligence. Nonetheless, it is appropriate to hold

[1] I am grateful for comments on this chapter to Terrell Carver, Karamvir Chadha, Jonathan Floyd, and two anonymous referees, and audiences at Balliol College, Oxford, and the University of Bristol. This article was inspired by an argument put to me by Andrew Williams. I am grateful to him for his generosity in allowing me to develop it here.

those relevantly connected to historic actions that would have been wrongful if performed with knowledge of their likely consequences liable for the costs of those consequences when a specific condition is fulfilled. This is when we are convinced that they would have acted as they did even if they had, in fact, been aware of the likely consequences of their actions. This claim will doubtless seem immediately counterintuitive to some: many have an understandable and deep-seated sense of indignation and resentment that is provoked when it is suggested that they might be blamed not for something they did but for something they would or might have done if they had the opportunity. The particular character of human history, specifically in relation to international affairs, and the precise sense of responsibility entailed in debates as to the modern-day allocation of costs associated with climate change mean that such a reaction is misplaced in relation to historic GHG emissions.

It is important to be clear at the outset as to the nature of the idea of responsibility of modern-day generations that is in play here. The conception invoked by the defender of backward-looking responsibility for the costs of climate change is not to be understood in terms of *moral* responsibility, which necessarily accrues to agents solely as a result of their actions. Given a commitment to a certain form of axiological individualism, which accepts that individuals rather than groups are the basic units of our ethical thinking, it follows that we can only be morally responsible for our own agency, not that of others. This is not to deny that we can act in ways that lead to our being members of groups that can bear collective moral responsibility for particular outcomes, nor that we can be responsible for our omissions as well as our actions, nor that attributions of moral responsibility can be appropriate in the absence of deliberate intentionality to bring about a particular outcome but in the presence of negligence or some other form of lack of care as to the possible effects of our agency. We cannot, however, be morally responsible for outcomes over which we had no kind of input or control, and that in no way result from choices that we have made. One way to state this point is to note that causal responsibility, understood in terms of being at least one causal factor in relation to an outcome, is a necessary though not sufficient condition for bearing moral responsibility for the outcome in question. By definition, we cannot be causally responsible for events that precede our existence, and so we cannot be morally responsible for them either. It does not follow from this, however, that we necessarily have no responsibility, in a different sense, for the effects of such events. David Miller helpfully distinguishes in this regard between ideas of moral and "remedial" responsibility. To bear remedial responsibility for a particular problematic outcome is to bear a special responsibility to correct the outcome, to act in such a way as to right whatever wrong or solve whatever problem has been created. Miller writes that "with remedial responsibility we begin with a state of affairs in need of remedy, ... and we then ask whether there is anyone whose responsibility it is to put that state of affairs right" (Miller, 2007: 98).

Neither moral nor causal responsibility is a necessary condition for possessing remedial responsibility for an outcome: trivially, I may possess remedial responsibility for another's actions if I promise that I will pay for the costs of what that agent does. Attributing remedial responsibility to an agent is therefore generally less significant than attributiing moral responsibility, since it does not entail any further claim about moral blameworthiness or the possible desirability of punishment. It merely holds that it is right for the agent in question to bear the costs of putting right whatever bad situation is at stake. Of course, there has to be some reason as to why an attribution of remedial responsibility is appropriate, as opposed to leaving the situation as it is and not shifting costs from the immediate victims to some other party. Moral responsibility would be an obvious reason for such a shift, but there are other possibilities, particularly in situations when agents with moral responsibility are unable to pay, such as in cases when they are dead or otherwise incapacitated. Thus Miller outlines what he calls his "connection theory" of remedial responsibility, whereby an agent can come to possess a duty to put a situation right on account of possessing one or more of six types of morally relevant connection to the bad outcome. Three of these are types of backward-looking responsibility: specifically, moral responsibility, outcome responsibility,[2] and causal responsibility. Miller then adds three further forms of connection: benefiting from the bad situation in question, possessing capacity to address the bad situation (e.g., being geographically well placed in relation to the victim, or being very wealthy and so being able to afford to assist), and possessing ties of community to those negatively affected. Which of these is the appropriate basis for allocating individual or shared responsibility in a given case will, he argues, inevitably depend upon our intuitively grounded reaction to the specifics of the situation. So he writes, "As far as I can see, there is no algorithm that could resolve such disputes. We have to rely on our intuitions about the relative importance of different sources of connection" (ibid.: 107).

When we apply this framework to contemporary policy questions relating to climate change, we find a complicated series of claims relating to where the costs associated with historic GHG emissions should fall. There are three broad categories of alternatives. The first is to do nothing, and leave the costs of climate change to lie where they fall, knowing that this will mean terrible hardship for some of the

[2] Outcome responsibility emerges, for Miller, when we ask whether "a particular agent can be credited or debited with a particular outcome – a gain or a loss, either to the agent herself or to other parties" (ibid.: 87). Miller distinguishes outcome responsibility from moral responsibility here since, on his account, to be morally responsible for X is to be liable for praise or blame in relation to X, and some outcomes are not appropriate subjects for this kind of moral appraisal (his examples include the effortless physical ability of a natural athlete and the disappointing horticulture of a clumsy gardener (cf. ibid: 89–90)).

world's most impoverished people. The second is to invoke some forward-looking principle, such as capacity to pay as a function of contemporary wealth, in order to allocate the costs of adaptation, mitigation, and compensation, as is done by reference to the "Ability-to-Pay Principle" (APP) (cf. Caney, 2010: 203–28; Meyer and Roser, 2010: 229–53). The third is to appeal to a backward-looking principle that links present-day parties, in one way or another, to particular historic emissions, either on its own or in combination with a capacity-oriented principle. The two most prominent attempts to do this are commonly known as the "Beneficiary Pays Principle" (BPP) and the "Polluter Pays Principle" (PPP) (cf. Butt, 2013). Both have well-established parallels in the theoretical literature on present-day responsibilities in relation to historic injustice. The first approach looks to the lasting benefits that historic industrial processes have had for people living in the present day. The most obvious modern-day beneficiaries in such cases may well be the descendants of those responsible for the original emissions, but this will be a contingent rather than a necessary relation: there may be descendants who have not so benefited, and there may be nondescendants who have (cf. Baatz, 2013: 94–110; Gosseries, 2004: 38–62; Page, 2011: 412–32; Page, 2012: 300–30). The second approach focuses more directly on the relation between ancestor and descendant, and posits some reason as to why it is appropriate for the latter camp to assume responsibility for the debts of the former (cf. Neumayer, 2000: 185–92; Pickering and Barry, 2012: 667–82; Shue, 1999: 531–45). A range of such reasons has been suggested in the historic injustice literature, ranging from accounts that place emphasis on the institutional continuity between the two groups through communal ties such as national membership or formal mechanisms such as the persistence of a state apparatus, to approaches that are grounded in the desirability of upholding transgenerational agreements or in the hypothetical wishes that would be expressed by previous generations if they were able to have their say in the present (cf. Butt, 2009; Spinner-Halev, 2012; Thompson J., 2002). In the cases of both the BPP and the PPP, it has been argued that the connection between past and present is of the right kind to place modern day parties under an obligation to try to put right the lasting effects of historical actors.[3]

What is crucial for the current argument is that whichever account is employed, it matters that the historic action in question was wrongful, that the parties responsible were culpable in moral terms. This is critical, since otherwise the claim that those responsible for historic emissions were not aware that they were doing

[3] Megan Blomfield has argued persuasively that to see backward-looking accounts of remedial responsibility for the costs of climate change as merely analogous to backward-looking accounts of remedial responsibility for the lasting effects of historic injustice is to mischaracterize the relation between the two and to miss the extent to which harm caused by the effects of climate change is often itself a result of past wrongdoing: so she argues, "the causal links between climate vulnerability and historic wrongs suggest that in some respects, the problem of climate change is actually part of an on-going, or enduring injustice" (Blomfield, 2015: 4).

anything wrong would not serve to block the claim for present-day remedial responsibility. The challenge results from the putative innocence of those responsible for GHG emissions prior to a specified date, which is taken to be the point at which it was reasonable to conclude that such emissions were morally problematic on account of the risk that they posed to the environment. Opinions vary as to when this was: while it is common to cite the first Intergovernmental Panel on Climate Change (IPCC) report in 1990, some have instead pointed to Svante Arrhenius's 1896 paper "On the Influence of Carbonic Acid in the Air upon the Temperature of the Ground" (Arrhenius, 1896). Axel Gosseries canvasses a range of different possibilities spanning more than 150 years, including 1840, 1896, 1967, 1990, and 1995, and of course, many continue to deny the effects of industrial processes on the climate, a mass of contrary scientific evidence notwithstanding (cf. Weart, 2008). Regardless of which point is chosen, the argument from excusable ignorance maintains that it is unfair to hold innocent emitters remedially responsible for the costs of their actions, as in Derek Bell's articulation of the principle, "If an agent is excusably ignorant of the consequences of her actions, she should not be held liable for the costs associated with the consequences of her actions" (Bell, 2011: 394). So, for example, Rudolf Schuessler writes that adopting a principle of strict liability in a context where those who pay the costs were not aware of the effects of their actions "comes close to primitive and/or totalitarian practices of clan liability" (Schuessler, 2011: 275), and Simon Caney has argued that if the ignorance of past emitters was excusable "it seems extremely harsh to make them pay for something that they could not have anticipated" (Caney, 2010b: 130–31). Imposing a duty on excusably ignorant emitters to benefit the victims of climate change is, he writes, "to prioritize the interests of the beneficiaries over those of the ascribed duty bearers. It is not sensitive to the fact that the alleged duty bearers could not have been expected to know. Its emphasis is wholly on the interests of the rights bearers and, as such, does not adequately accommodate the duty-bearer perspective" (ibid.: 131–32).

Broadly speaking, there are three ways to challenge the argument that present-day parties do not owe remedial duties for the effects of historic emissions as a result of the nonculpable ignorance of the emitters. The first is to question the claim that such emissions were indeed made in a context of nonculpable ignorance. This can be done by maintaining that (at least some) past generations should have been aware that their actions either were harmful or posed a morally significant degree of risk, meaning that the emissions in question either were straightforwardly wrong as harmful or were culpably negligent. The second is to maintain that the moral character of those responsible for the emissions is not relevant to modern-day assignations of remedial responsibility, on the grounds that historic emitters were remedially responsible for their actions even in the absence of moral culpability, meaning that modern-day parties who are linked to them in a morally relevant way (as in the PPP) or who have benefited from their actions (in relation to the BPP) are liable for the associated lasting costs. A recent article by Alexa Zellentin, for example, makes both claims. Zellentin (2014) first challenges the idea that GHG emissions associated with the Industrial Revolution really were innocent, arguing that there was at least "some indication that industrialization posed environmental threats, and the relevant agents could therefore have been expected to know that they were embarking on a path of action a) that was not fully under their control, and b) the impact of which was not foreseeable" (ibid.: 270).[4] She then argues that rectificatory duties can arise in the absence of blameworthiness when an agent infringes the rights of others, writing that "responsibility and blameworthiness can come apart. This justifies the imposition of rectificatory duties on those who have infringed the rights of others even where the infringement happened in a non-blameworthy manner" (ibid.: 261).[5]

I am sympathetic to both of these responses, but the current analysis sets them to one side. This leads to the third form of response. This accepts, if only for the sake of the argument, that the emitters cannot be said to be morally culpable in relation to the specific emissions they produced, as they were ignorant of the likely effects of the emissions in question. It maintains, however, that more advanced scientific knowledge would have made no difference to the actions of the emitters or to those responsible for regulating them: even if they had known of the likely consequences of their actions, they would have carried on regardless. It therefore maintains that the exculpatory strategy does not block the transmission of remedial responsibilities to agents in the present day. This does not amount to a general argument in favor of strict liability for the effects of emissions, as suggested by the second strategy: instead, the claim is that causal responsibility of this kind leads to remedial responsibility only in a specific subset of cases when parties bring about a particular outcome: that which obtains when a counterfactual condition is satisfied that attests to the nonrelevance, in practical terms, of the ignorance of the causally responsible agents, who would have acted in the same way even if they had known of the likely effects of their actions.[6]

[4] See also Meyer (2004: 20–35) for an argument relating to the effects of "wrongless historical emissions" that lead current generations to fall below a threshold notion of harm, and Heyward (2014: 405–19).

[5] See also Bell (2011).

[6] This idea of "likely effects" needs unpacking. The idea here is not to imagine that past generations had perfect clairvoyance as to how the future development of the world would play out, not least as such perfect knowledge would extend to contemporary debates over how to allocate the costs of climate change, and indeed to the future resolution, if any, of these debates. Rather, the question is how we believe past generations would have acted had they had at least some sense as to the likely future impact of their actions, and some understanding of the threat they were posing to future generations, even if we accept that they would have been unsure whether that threat would in fact materialize. Of course, even in the present day, scientists are unsure as to how existing levels of GHG emissions will affect future generations, and so climate policy discussions typically proceed with

Such a claim faces two challenges: one epistemic and one normative. The epistemic challenge queries whether we can know how historic agents, or indeed any agents, would have acted had they been in possession of information that would have rendered their actions, had they then continued with them, immoral. The normative challenge disputes the moral relevance of the counterfactual condition: even if we accept that agents would have acted in this way, it argues, it makes no difference to the remedial responsibilities of either the agents or subsequent related parties. We will consider each in turn.

First, how can we know how people would have acted had they been in possession of information that they did not, in fact, possess? The obvious way to respond is to say that we cannot. It is true that we cannot say for certain how given agents would have acted had circumstances been different. My claim is that the relevant standard for the purposes of this chapter, however, is not certainty. It is sufficient to say that we are convinced that agents would have acted in the same way regardless of the harmful effects of their actions, that it seems highly or overwhelmingly likely that this would have been the case. This idea can be expressed in a number of ways. We might, for example, borrow the standard of proof used in criminal trials, in which jurors are required to be persuaded of the guilt of a defendant "beyond reasonable doubt," as opposed to the standard often employed in civil trials where cases are instead settled on the balance of probabilities. Is this condition satisfied in relation to historic emissions – is it the case that we should be convinced, beyond a reasonable doubt, that the great bulk of historic emissions would have been produced even if previous generations had been aware of their possible future effects? It seems to me reasonably clear that this is indeed the case. Two observations, in particular, may be made in relation to the claim. The first concerns the general character of industrial powers historically, particularly in relation to their dealings with nonnationals. The dominant mode of foreign policy for much of the period in question was imperial and/or colonial: many less developed countries were subject to grievous wrongdoing at the hands of Western countries eager to fuel their industrial growth. It stretches credibility to believe that a concern for the descendants of those who were being dominated and subjugated, in particular, would have checked the progress of the Industrial Revolution. It is not hard to point to myriad policies that were clearly harmful to others but were justified by crude invocations of the national interest, with a particular emphasis on the interests of

a number of different scenarios in mind (cf. UNEP 2011). Understanding the "likely consequences" of their actions, then should be understood in terms of describing a state of affairs in which, to repeat Zellentin's words, agents knew at least "that they were embarking on a path of action a) that was not fully under their control, and b) the impact of which was not foreseeable" (Zellentin, 2014: 270), and further that some of the possible scenarios associated with their actions involved very substantial future environmental harm.

those living at the time. Second, and perhaps more straightforwardly, it seems clear that the development of knowledge about the effects of GHG emissions has not in fact served to check emissions in any kind of meaningful way. As Henry Shue (1999: 536) writes:

> The industrial states' contributions to global warming have continued unabated long since it became impossible to plead ignorance. It would have been conceivable that as soon as evidence began to accumulate that industrial activity was having a dangerous environmental effect, the industrial states would have adopted a conservative or even cautious policy of cutting back greenhouse emissions or at least slowing their rate of increase. For the most part, this has not happened.

Stephen Gardiner argues even the most high-profile attempts at action, the Kyoto Protocol and the Copenhagen Accord, have been ineffectual, observing that the aftermath of the Kyoto renegotiation in 2001 actually saw widespread increases in emissions in the industrialized world as a whole (Gardiner, 2011: 132). His verdict on these initiatives is damning: "In essence, it is highly plausible to believe that, at best, these efforts tried only to do something limited to protect the interests of the present generation, narrowly defined, and that, at worst they served merely as cover for business as usual. Either way, they did almost nothing to aid future generations, or the planet more generally" (ibid.: 128).[7] Perhaps, optimistically, we are now seeing some signs of progress, in at least some parts of the world. But this is limited, it is uncertain, it takes place in an international arena increasingly characterized by the spread and consolidation of international law, and it has, to put it mildly, been a long time coming.

Putting both these observations together creates a convincing case for the claim that the counterfactual condition is satisfied in relation to historic emissions. Perhaps neither on its own is sufficient to establish the claim: the fact that X does not change course when presented with evidence that what she is doing is wrong may not be in itself sufficient to establish the stronger claim that X would have set out on this course in the first place had she known of the effects of X, given the significance of path dependence.[8] When we look, however, at the moral character of historic communities along with the evidence of how these communities reacted when they became aware of the effects of their actions, the case becomes compelling. It is important to note here that we are thinking about the results of the aggregated actions of large numbers of individuals, along with the

[7] (Cf. Butt, 2016).

[8] It is worth noting, however, in contrast to the case of carbon emissions, that we do have experience of swift and timely international action in light of new scientific knowledge in order to combat potentially catastrophic environmental change, as witnessed by the international community's response to revelations about the role of chlorofluorocarbons (CFCs) in the depletion of the ozone layer in the 1987 Montreal Protocol. The contrast with the role of the international community in relation to GHGs is striking.

ongoing policy decisions of those responsible for regulating them (or choosing not to so act). The counterfactual condition in this context would not typically be derailed by one individual, or a small number of individuals, choosing to act differently: the claim of this section is that the overall picture would have been so similar as to be practically indistinguishable from the situation we face in the present day. I shall not labor the point further. In my view, it is, to say the least, very hard to make a good faith argument that matters would have been different had scientific knowledge been more advanced at an earlier date. We should indeed accept that we cannot know for certain what would have happened, but it is hard for me to accept that there is much doubt about the most likely outcome. There may be legitimate disagreement here about how far beyond the balance of probabilities we must progress for the counterfactual condition for obtain, but insisting upon certainty seems too demanding a standard. The danger is that to make such an argument – or, more likely, to insist that one's opponent does the impossible and prove that a different outcome would not have been the case – will frequently be to argue in bad faith, from a motivation of seeking to avoid liability in the present rather than of trying to derive a fair-minded assessment of the character of historic policy making. Even if one believes that it is legitimate for lawyers to seek to protect their clients in such a fashion, it does not follow that it is a justifiable tactic for democratic communities that are struggling with the question of what they owe to others.

So suppose we accept all of the preceding argument. We agree (given this chapter's assumptions) that had previous generations acted in knowledge of the likely effects of GHG emissions, then they would have acted wrongfully, and that this would have given rise to modern-day remedial obligations on the part of those relevantly connected. We also accept that accurate scientific knowledge would not, in fact, have made any difference to those actions. Why should this latter acknowledgment give rise to remedial obligations in the present day? To show why this might be thought to be the case, we need to return to the earlier distinction between moral responsibility, on the one hand, and remedial responsibility, on the other hand. Different understandings of responsibility are useful in different contexts. This applies, for example, when we think about how the law deals with ideas of responsibility. When it comes to criminal responsibility, it is a necessary condition for finding agents guilty of crimes that they are morally responsible for the commission of the crimes in question. Whichever theory of criminal justice we endorse, and however we justify using the coercive power of the law to impose hard treatment on criminals as punishment, a guilty verdict is predicated upon some sense of blame: the idea that the criminal did something wrong. The amount of punishment imposed need not be in exact proportion to the crime committed: thus if multiple individuals are found guilty of plotting to commit the same crime, their punishment is not normally a proportionate share of the punishment that an individual who was convicted of perpetrating the same crime on his own would face, even if the sentence given in the latter case is higher than the average sentence given the coconspirators in the former case. The quantity of punishment, and whether, indeed, anyone is punished at all, depend on whether there are agents who have acted in a culpable way. It is crucially important in such cases that the proper relation between the agent and the crime should exist: we rightly think it would be wrong to punish people for something they did not do, or that they did innocently. So in criminal law, "*mens rea,*" a guilty mind, is a strictly necessary condition for a conviction. The guilt that is relevant here pertains to the specific crime in question: it is clearly not (or should not be) enough for a prosecutor to point to the general bad character of the accused. We instinctively recoil from the kind of legal system proposed in the story and film *Minority Report* (originally written by Philip K. Dick) in which predictions are made as to which crimes individuals will commit, and the would-be perpetrators are apprehended and subsequently punished before they do in fact commit the crime. People should not be punished for things they did not do, regardless of how likely it is that they would have done them. This does not, of course, mean that they cannot be punished for conspiring or attempting to commit a crime – it can be legitimate to intervene to prevent wrongful action, and it can be legitimate to punish the preparation for wrongdoing even if the perpetrator does not actually have the chance to perpetrate. But there is nothing to be gained and much to be lost, from a moral perspective, in punishing the innocent.

The kind of case under consideration in this chapter, however, is not quite the *Minority Report* case. This is not a situation when an act is prevented and a counterfactual judgment is made as to what would have happened had the intervention not occurred. Instead, an act does occur, and the counterfactual concerns our judgment as to whether the act would have occurred in the same way had the perpetrators been aware of the likely effects of their actions. It is true that we make a moral appraisal of the character of perpetrators in such judgments. However, and crucially, this is not an appraisal made with a view to deciding whether the perpetrators should be punished. Instead, the point of considering the moral character of the action is to determine not moral but remedial responsibility. This distinction is key. In such a case, it is straightforwardly true that the perpetrators are causally responsible for the effects of the action under consideration. It is also, by hypothesis, true that they are not morally responsible for the effects of the action, as they are nonculpably ignorant as to what these effects are likely to be. The question is, in Miller's terminology, whether we should also hold the perpetrators remedially responsible for the action: whether the costs of the actions can justifiably be laid at their door. My instinct is that the burden is on the perpetrators to show why the costs of their actions should not be borne by them but should be left to lie where they fall, or should be borne by some other agent or agents. The situation in this case has the character of a zero-sum game – costs have been accrued, and someone has to bear them. Leaving costs to fall where they lie is a principle with some prima facie plausibility: there is administrative cost to shifting burdens, and so all else being equal, there is a reason not to intervene. This is not, however, a terribly strong principle. Thus we have seen that some wish to operate a principle of strict liability in relation to historic emissions, and to hold those who were outcome responsible also remedially responsible regardless of their moral culpability. If this claim is to be resisted, it must be maintained that the innocence of the perpetrators provides a sufficiently good reason not to make such a claim, and instead to argue that the perpetrators have no relation to the costs of the harm that is sufficiently strong to trump either a forward-looking principle of capacity or equal sharing, or the leaving harms where they lie principle. My argument in this section is that satisfaction of the counterfactual condition relating to likely consequences tips the balance against the perpetrators. It significantly increases the moral relevance of their connection to the wrongdoing beyond mere causal responsibility and means that potentially those with greater capacity, but certainly the victims of their actions, have justifiable cause for complaint if they, rather than the perpetrators, are forced to bear the cost of the perpetrators' actions. My claim here is that the default situation in such a case is not that of leaving the harm to lie where it falls, but of shifting it to the perpetrators. We can be justified in reversing this situation, and not only exculpating but indemnifying the perpetrators, but only if the innocence of the perpetrators goes, we might say, all the way down. It is not hard to think of cases when actions have truly bizarre consequences, and when we are generally happy to say that ignorance is a defense not only against moral blame but potentially against remedial responsibility. Consider first the following example, taken from Judith Jarvis Thomson (1990: 229):

DAY'S END: B always comes home at 9:00 pm and the first thing he does is to flip the light switch in his hallway. He did so this evening. B's flipping the switch caused a circuit to close. By virtue of an extraordinary series of coincidences, unpredictable in advance by anybody, the circuit's closing caused a release of electricity (a small lightning flash) in A's house next door. Unluckily, A was in its path and was therefore badly burned.[9]

Morally speaking, there does not seem to be any significant difference between this case and one that involves some kind of "butterfly effect": it is hard to see that anything of any moral consequence ensues if I clap my hands and this inadvertently causes a tornado on the other side of the world. In this situation

[9] I am grateful to Mike Otsuka for this reference. For discussion, see Tadros (2011: 220).

the harms seem to be nothing more than an unfortunate accident, and so presumably they should be treated as such under our preferred account of distributive justice, and so, for example, be left to lie where they fall, or be met by the community as a whole: either way, there seems no compelling reason to make B pay. The more direct the causal link, the less comfortable we may feel about such an outcome, but the intuition has force even when one acts innocently with disastrous effects to those directly around one: such cases are tragic and unfortunate, but if it really is true that the effects in question were genuinely unforeseeable and that there is no question of culpable negligence, there is certainly a case for saying that they should be treated as if they were natural disasters rather than the results of human agency for the purposes of assigning remedial responsibilities. The perpetrator is inextricably linked to the outcome, but she can distance herself from it by stressing the lack of culpable intent on her part. Very often, she will do this explicitly by stressing her ignorance of the effects of the action and stressing that had she known what was going to happen, she would not have acted in the way she did. ("I am so sorry. I had no idea that would happen. Obviously I would never have acted in this way had I known what would have happened – you must believe me!")

Some situations, however, are much less straightforward. Consider the following case:

PESTICIDE: A corporation manufactures and sells a popular pesticide. There are two ways to produce this pesticide, involving either substance X or substance Y. The substances cost the same, and so the corporation utilizes both in different batches, simply purchasing whichever it can most easily obtain each time it restocks its supplies. The corporation becomes aware that substance X causes long-term negative side effects to people who are exposed to it via food consumption; however, extensive testing has shown no such effects for substance Y. The corporation does not exclude substance X from its production, but continues to use X and Y indiscriminately in different batches, depending on whichever substance can most easily be obtained. As a result, many people who eat food treated with pesticide from the substance X batch develop significant health problems. It then transpires that substance Y, contrary to the best scientific information of the time, causes similar health problems of a comparable magnitude.

It seems clear that those negatively affected by substance X have been wrongfully harmed by the corporation, and that the corporation is both morally and remedially responsible for the harm they have suffered. What of those affected by substance Y? There seems to be no moral responsibility of a straightforward kind in this case if, by stipulation, the corporation was not negligent in their use of Y. In a criminal trial, those responsible for the corporation's actions could only be convicted in relation to the harms caused by substance X. The absence of *mens rea* would be an adequate defense in relation to harm suffered as a result of substance Y. My claim, however, is that the remedial responsibility of the corporation in this case is importantly different from that of a corporation who knew of the effects of X but not of Y, and so for that reason only used Y in their products. The second corporation is innocent of wrongdoing. Some, of course, will argue in favor of a principle of strict liability in such a case, and hold that it should be held remedially responsible regardless of the absence of wrongdoing. It does, though, seem significantly more justifiable to impose the costs associated with its use of substance Y on the first corporation. This corporation is not in good moral standing. It has made clear by its actions that it does not view the well-being of those it affects as a significant concern. It is true that the corporation can truthfully assert to those affected by Y that it did not know its actions would cause them harm. It seems, however, that it cannot in good conscience say that it would not have distributed the pesticide had it known of its effects: its other actions rob such a claim of its plausibility, and the exculpatory defense of ignorance cannot be employed in good faith. The counterfactual condition, then, seems clearly to be fulfilled: the corporation treated X and Y identically even though it believed them to have significantly different effects, so it strains credibility to argue that it would have acted differently had it learned of the true nature of Y at the same time as it properly understood X. On my account, this means that its connection with the harms caused by Y is such that it is appropriate to hold the corporation remedially responsible. We need not insist that the cases of the two groups of victims are identical from the perspective of reparative justice; in particular, one can imagine a coherent argument that would seek to employ a forward-looking perspective focusing on, for example, capacity in relation to harms caused by substance Y while insisting that remedial responsibility be borne by the perpetrator in response to substance X. What would be flatly unacceptable, on my view, however, would be to assign remedial responsibility to the corporation in relation to the substance X cases but do nothing about the Y cases, leaving the harm to lie where it falls, on the victim. Imagine that there is no prospect of assistance from a third party. In such a case, with no other agent in a position to improve the condition of the Y victims, the causal responsibility of the corporation along with the satisfaction of the counterfactual condition are sufficient to give rise to remedial responsibility.[10]

[10] I have presented causal responsibility and the satisfaction of the counterfactual condition as jointly sufficient to give rise to remedial responsibility. Are both also jointly necessary conditions in circumstances of harmful conduct perpetrated in ignorance of its effects? This is slightly complicated. Satisfaction of the counterfactual condition has been presented as necessary given the chapter's commitment to accepting the prima facie force of the exculpatory block, though as noted previously, I am nonetheless in fact sympathetic to the strict liability model: my purpose here, however, is to maintain that those who resist strict liability on grounds of historical ignorance should exempt cases when the counterfactual condition is satisfied. For the most part, I also believe we should see causal responsibility as necessary for an ascription of remedial responsibility stemming from the counterfactual condition. The fact that actual outcomes, in this case, particular historical emissions, have actually come about matters here.

Suppose, then, that the preceding argument is accepted. This is not quite enough to make the case for present-day remedial responsibilities as the result of historic emissions, but there is not much more work to be done. Recall that it was accepted for the sake of argument that it is the employment of the exculpatory block, and only the exculpatory block, that means that present-day parties do not possess remedial responsibilities as a result of historic emissions. This being the case, it is accepted that those emissions that were morally problematic did historically give rise to remedial responsibilities, on account of the link of present-day parties to the emissions in question, whether this be on account of a present-day link to the lasting effects of the emission through the "beneficiary pays" principle, or to the emitters, on account of the "polluter pays" principle. So we can now fit the historic emissions that did give rise to historic remedial responsibility on account of the counterfactual condition into the contemporary picture. In terms of the BPP, we now have modern-day benefits that have arisen as a result of historic actions that were morally problematic in the sense that they caused harm to others and would have been performed even had the emitters been aware of their consequences. In terms of the PPP, we have historic emitters whose moral innocence cannot be demonstrated "all the way down," and who bear a morally relevant form of connection to present-day actors. This is sufficient to tip the balance, meaning that historic emissions of this type should be treated as if they were wrongful.

I have presented the argument thus far in a particular form, whereby the counterfactual condition makes the crucial difference between a situation when the causally responsible agent does or does not possess remedial responsibilities. The real world context of climate change policy is, however, a little more complicated than this would suggest. The principle of "Common but Differentiated Responsibility" (CBDR) formulated in Principle 7 of the Rio Declaration articulates the sense in which a multiplicity of actors has responsibilities for the costs of mitigation, adaptation, and compensation resulting from climate change.[11] Some

We are not asking who *deserves* to pay certain costs, if this is to be understood in terms of asking who is morally deficient in terms of displaying a moral character such that they would have acted wrongly had circumstances allowed them to do so. It may well be that we conclude that many historic agents would have produced substantial emissions had they been in a place to do so, regardless of their knowledge of their possible effects, just as we might believe that various historic peoples would have engaged in colonial injustice had they had the capacity to do so. The chapter does not argue that counterfactual wrongdoing should be punished, but rather that the costs of actions should stay with those who caused the actions in question when they would have acted as they did even if they had not been ignorant. It is possible that in some very particular cases one might wish to argue that the counterfactual condition on its own forms a morally relevant form of connection to an action such that an ascription of remedial responsibility would be appropriate, but this point need not be pressed for present purposes.

[11] "In view of the different contributions to global environmental degradation, States have common but differentiated responsibilities. The developed countries acknowledge the responsibility that they bear in the international pursuit of sustainable development in view of the pressures their societies place on the global environment and of the technologies and financial resources they command" (UNCED, 1992). For discussion, see Harris (1999:27–48).

of these responsibilities stem from backward-looking concerns, rooted in principles such as the BPP and the PPP; others are focused on forward-looking considerations such as capacity as in the APP. The question here is not really one of identifying a particular agent or agents who should bear the costs of climate change while others are let off the hook: instead, the real question is how much remedial responsibility should be apportioned to different actors, and how much of the cost should be left to fall where it lies. It may, then, be misleading to think about the counterfactual condition as a mechanism for attributing remedial responsibility to an agent who would otherwise get off scot-free when we come to think of the specific context of climate change. Instead, the context is one where there is a multiplicity of different arguments that link particular industrialized democracies to the lasting effects of GHG emissions. The responsibilities of some states are overdetermined, as they score highly in terms of historic responsibility, modern-day benefit, and modern-day capacity; those of others depend upon the amount of weight placed on different morally relevant forms of connection. So one way to interpret the argument of this chapter would be to see it as seeking to add extra weight to backward-looking principles such as BPP and PPP in this debate: from this perspective, no one would suddenly acquire remedial obligations simply on account of the counterfactual condition, but rather, the satisfaction of the counterfactual condition might be considered alongside other arguments linking past and present.

The strong claim made in this chapter, that those who bear moral responsibility can possess the same degree of remedial responsibility as those who bear mere causal responsibility but also satisfy the counterfactual condition, is undoubtedly controversial. By way of conclusion, it may be helpful to note ways in which this strong claim may be qualified for those who accept that there is some moral issue at stake in contexts where the counterfactual condition is met, but who are not willing to accept the full claim in this unambiguous form. First, it might be noted that the counterfactual condition is significant if it is taken only to have the effect of lessening the importance of, rather than strictly negating, the exculpatory block. It is quite possible to maintain that an agent who satisfies the counterfactual condition in relation to ignorant harming has a degree of remedial responsibility, if not as much as an agent who bears full moral responsibility for a wrongful harm; similarly, one might hold that satisfying the counterfactual condition would increase the strength of the moral reasons an agent has to assume remedial responsibilities, rather than maintaining that such responsibility strictly holds when the condition is satisfied and does not hold when it is not. This can be seen in relation to the "beneficiary pays" principle. I have argued elsewhere that the BPP is potentially capable of encompassing a variety of different contexts where a party involuntarily receives benefits stemming from the actions of another (Butt, 2014: 336–48). At one extreme, we might imagine benefits that result directly from unambiguous wrongdoing, the intention of which was precisely to improve the material situation of the involuntary beneficiary in question. At the other, we might imagine cases when the agents who suffer a loss are responsible to some degree for the loss that they have suffered, as when they choose to provide a service to another without that other's prior consent in the hope of receiving future payment. I have argued that there is a case in each situation to be made for the claim that the beneficiary can possess good moral reasons to offset the loss suffered by the other party, so long as doing so does not leave the beneficiary worse off in overall terms than he or she would have been had the act in question not occurred. The strength of the moral reasons will, however, be rather different in the varying cases. I have argued that in the case of wrongful benefits designed to aid the beneficiary at another's expense these reasons are particularly strong, and mean that a beneficiary who does not act but is content to enjoy the benefit is clearly blameworthy. By contrast, it would not be unreasonable to suppose that the kind of reasons associated with paying for unsolicited benefits are generally supererogatory in moral terms, whereby it may be good from a moral perspective to pay, but whereby one does nothing wrong and should incur no blame for not so acting. The case of purely accidental harms and benefits falls somewhere between these two cases. There is particular reason to act in the case of wrongdoing that stems from a condemnation of the blameworthy actions of the perpetrators: such a concern is generally missing in cases of innocent harm doing that does not involve culpable negligence. My claim is that the satisfaction of the counterfactual condition changes the picture here: cases with this character look rather more like wrongful harm, and rather less like innocent happenstance. When modern-day benefits arise from the actions of forebears who were generally morally problematic, we have good reason to seek to distance ourselves from the lasting effects of their harmful actions, whether anticipated or not, by redressing the harm they have caused.

Second, it is possible (though not, it should be stressed, in my view necessary) to concede that the kind of remedial duties that one acquires in relation to the counterfactual condition are not enforceable obligations, which can be upheld by third parties under the threat of coercive force, but are instead moral duties, which should be reflected upon and acted upon by present-day parties who wish to act with moral integrity in relation to others. From this perspective, what modern-day parties should be asking themselves is not the legalistic question of what the minimum is that they can get away with paying as a result of the past actions of their forebears and the provenance of their modern-day advantages. Instead, both political leaders and democratic publics should seek to determine what form of contemporary action is the proper response to their own particular historical context. Too often, in the broad field of international reparative justice, states seek to pursue strategies of the former kind, nakedly rooted in self-advantage rather than pursuing any kind of good faith effort to determine what they owe to others. Consider, for example, the initial response of the UK government in 2011 to claims for compensation made by Kenyan victims of torture and sexual abuse during the Mau Mau uprising of the 1950s, prior to agreeing to pay limited compensation in 2013. The UK government at first sought to deny the claims on two grounds: first, that the statute of limitations for such actions had passed; and second, that responsibility for the actions of the British colonial administration passed to the nascent Kenyan government at the end of the decolonization process in 1963 (cf. Bowcott, 2011). Both arguments are woeful from a moral perspective. There is a range of good practical reasons why some crimes should be subject to statutes of limitation in law, especially, perhaps, associated with the desirability of individuals being able to live their lives free from the uncertain prospect of possible litigation, but such concerns cannot have significant moral force when invoked by institutions such as states in contexts where it is abundantly clear that earlier attempts to bring suit would have been doomed to failure. Similarly, there are obvious reasons why the terms given to an oppressed people as the condition for their liberation may be suspect from a moral point of view and should not necessarily be seen as subsequently enforceable. These are the arguments of a lawyer seeking to avoid liability, not of a political actor possessed of a genuine desire to see that justice is done in relation to the past. Similarly, there is a real risk of grotesquely bad faith argumentation if modern-day states seek to evade responsibility for their historic emissions on technical terms relating to historic ignorance, in a context where it is abundantly clear that this ignorance was not, in fact, action guiding. Whether the duties that arise under the principle discussed in this chapter could theoretically be justifiably enforced can be left as an open question if the answer is thought to be problematic; what should be accepted, on my view, is the claim that political communities who wish to present themselves as moral actors on the world stage should not seek to hide behind the ignorance of their forebears when it is clear that this ignorance had no material impact on outcomes. Confronting the legacy of the past is a painful business, and particularly so when it comes to being honest about not only the cost at which our present-day advantages came about but also the moral character of those whose actions gave rise to the advantages in question.

References

Arrhenius, S. (1896). On the Influence of Carbonic Acid in the Air Upon the Temperature of the Ground. *The London, Edinburgh, and Dublin Philosophical Magazine and Journal of Science*, **41**, 237–76.

Baatz, C. (2013). Responsibility for the Past? Some Thoughts on Compensating Those Vulnerable to Climate Change in Developing Countries. *Ethics, Policy & Environment*, **16**, 94–110.

Bell, D. (2011). Global Climate Justice, Historic Emissions, and Excusable Ignorance. *The Monist*, **94**, 391–411.

Blomfield, M. (2015). Climate Change and the Moral Significance of Historical Injustice in Natural Resource Governance. In *The Ethics of Climate Governance*, ed. A. Maltais and C. McKinnon. London: Rowman and Littlefield, pp. 3–22.

Bowcott, O. (2011). Mau Mau Victims Seek Compensation from UK for Alleged Torture. URL: http://www.theguardian.com/world/2011/apr/07/kenyans-mau-mau-compensation-case.

Butt, D. (2009). *Rectifying International Injustice: Principles of Compensation and Restitution Between Nations*. Oxford: Oxford University Press.

Butt, D. (2013). 'The Polluter Pays': Backward-Looking Principles of Intergenerational Justice and the Environment. In *Spheres of Global Justice*, ed. J.-C. Merle. Dortrecht: Springer.

Butt, D. (2014). 'A Doctrine Quite New and Altogether Untenable': Defending the Beneficiary Pays Principle. *Journal of Applied Philosophy*, **31**, 336–48.

Butt, D. (2016). Law, Governance and the Ecological Ethos. In *The Oxford Handbook of Environmental Ethics*, ed. S. M. Gardiner and A. Thompson. Oxford: Oxford University Press.

Caney, S. (2010a). Climate Change and the Duties of the Advantaged. *Critical Review of International Social and Political Philosophy*, **13**, 203–28.

Caney, S. (2010b). Cosmopolitan Justice, Responsibility, and Global Climate Change. In *Climate Ethics: Essential Readings*, ed. S. M. Gardiner, S. Caney, D. Jamieson, H. Shue. Oxford: Oxford University Press, pp. 122–45.

Gardiner, S. M. (2011). *A Perfect Storm: The Ethical Tragedy of Climate Change*. Oxford: Oxford University Press.

Gosseries, A. (2004). Historical Emissions and Free-Riding. *Ethical Perspectives*, **11**, 38–62.

Harris, P. G. (1999). Common but Differentiated Responsibility: The Kyoto Protocol and United States Policy. *Environmental Law Journal*, **7**, 27–48.

Heyward, C. (2014). Benefiting from Climate Geoengineering and Corresponding Remedial Duties: The Case of Unforeseeable Harms. *Journal of Applied Philosophy*, **31**, 405–19.

Meyer, L. H. (2004). Compensating Wrongless Historical Emissions of Greenhouse Gases. *Ethical Perspectives*, **11**, 20–35.

Meyer, L. H. and Roser, D. (2010). Climate Justice and Historical Emissions. *Critical Review of International Social and Political Philosophy*, **13**, 229–53.

Miller, D. (2007). *National Responsibility and Global Justice*. Oxford: Oxford University Press.

Neumayer, E. (2000). In Defence of Historical Accountability for Greenhouse Emissions. *Ecological Economics*, 33, 185–92.

Page, E. (2011). Climatic Justice and the Fair Distribution of Atmospheric Burdens. *The Monist*, 94, 412–32.

Page, E. (2012). Give It Up for Climate Change: A Defence of the Beneficiary Pays Principle. *International Theory*, **4**, 300–30.

Pickering, J. and Barry, C. (2012). On the Concept of Climate Debt: Its Moral and Political Value. *Critical Review of International Social and Political Philosophy*, **15**, 667–85.

Schuessler, R. (2011). Climate Justice: A Question of Historic Responsibility? *Journal of Global Ethics*, **7**, 261–78.

Shue, H. (1999). Global Environment and International Inequality. *International Affairs*, **79**, 531–45.

Spinner-Halev, J. (2012). *Enduring Injustice*. Cambridge: Cambridge University Press.

Tadros, V. (2011). *The Ends of Harm: The Moral Foundations of Criminal Law*. Oxford: Oxford University Press.

Thompson, J. (2002). *Taking Responsibility for the Past: Reparations and Historical Injustice*. Cambridge: Polity.

Thomson, J. J. (1990). *The Realm of Rights*. Cambridge, MA: Harvard University Press.

United Nations Environment Programme (UNEP) (2011). *Bridging the Emissions Gap: A UNEP Synthesis Report*. URL: http://www.unep.org/pdf/UNEP_bridging_gap.pdf. Last accessed December 29, 2015.

United Nations Conference on Environment and Development (UNCED) (1992). *The Rio Declaration on Environment and Development*. URL: http://www.unep.org/Documents.Multilingual/Default.asp?documentid=78&articleid=1163.

Weart, S. R. (2008). *The Discovery of Global Warming*. Cambridge, MA: Harvard University Press.

Zellentin, A. (2014). Compensation for Historical Emissions and Excusable Ignorance. *Journal of Applied Philosophy*, **32**(3), 258–74.

4

How Legal Systems Deal with Issues of Responsibility for Past Harmful Behavior

DANIEL A. FARBER

I Introduction

Because carbon dioxide has a long lifetime in the atmosphere, climate damage is a function of past emissions, not just current ones. Thus, developed countries such as the United States have created a disproportionate amount of atmospheric carbon dioxide compared to their present emissions or projected future emissions. This disparity gives rise to claims by victims of climate change in developing countries for recompense (Farber, 2008). Such recompense could take the form of damage awards or funding for adaptation or mitigation efforts. Alternatively, historic emissions could be taken into account in allocating responsibility for future emission reductions. A particularly relevant issue is the extent to which liability can be based on past conduct that took place before the severity of the risk was apparent.

In sorting through the competing claims about historic responsibility, consideration of analogous legal rules can be relevant for three purposes. Most obviously, it could be relevant to considering the possible limits of legal liability for greenhouse gas emissions. Although this chapter will not directly focus on this issue, which has been extensively studied elsewhere, courts considering the novel liability issues relating to climate change could well be influenced by the legal system's treatment of similar types of issues.

Further, consideration of existing liability regimes could also provide a gauge of the likely willingness of developed countries to accept responsibility for past carbon emissions in negotiating a climate agreement. Clearly, the scale of the issues is much different, and nations might feel differently about responsibility for harm to other nations versus responsibility for harm to their own citizens. Still, proposals for international greenhouse gas responsibility are likely to meet increasingly great difficulties to the extent they extend beyond even what nations consider appropriate in dealing with analogous domestic issues. Consideration of existing legal regimes might also be relevant to arguments about whether historic responsibility for past emissions could be implemented feasibly. The existence of successful liability schemes that impose similar forms of responsibility in other contexts might be relevant to such arguments, although again differences in scale would have to be taken into account.

Finally, in considering the moral issues, it may be useful to understand the considered judgments of judges and democratic leaders in deliberating on analogous problems. Laws are not purely judgments about social justice and can be influenced by other factors such as interest group influence or public overreactions to problems. Thus, it would be a mistake to take them as proof of considered moral judgments. Even to the extent they are, they may reflect only the views of certain segments of society, or they might be distorted by various biases. Theories of ethics may vary as to what role, if any, they give moral intuitions or ordinary moral deliberation. Yet, on many theories, considered moral judgments do serve at least some role as inputs for ethical argument. Existing legal rules, to the extent they do reflect society's moral views, may therefore be relevant.

Section II analyzes liability for environmental harm in international law. Although international law is underdeveloped compared with domestic law, it is particularly relevant for the obvious reason that climate change is a global problem. To the extent that states have accepted forms of historic responsibility toward each other for environmental harm as a general principle, there is a clear argument that climate change should be no exception.

The following two sections of the chapter, sections III and IV, focus on the United States and the European Union. The analysis focuses on these legal systems for three reasons. The first is that their legal systems have been influential globally, in part because they have geopolitical power, in part because they confronted the legal challenges of industrialization earlier than other countries, and in part because they were considered more "advanced." Second, it is not feasible in a short chapter to analyze all of the world's legal systems, or even to give detailed attention to each of the legal regimes adopted by American states or every member state of the European Union. It is only possible to comment on general trends in some leading legal systems. The United States and the European Union were among the first to address environmental issues legally. Third, because the European Union and the United States are the leading sources of historic emissions, their treatment of analogous legal problems is particularly relevant.

In considering the EU and U.S. legal regimes, this chapter focuses primarily on analyzing legal settings where past acts can cause harm far in the future – products liability and hazardous waste – in order to examine how changes in technology and scientific knowledge affect liability. Because carbon dioxide remains in the atmosphere for such long periods, current concentrations are the result of emissions from the heyday of the Industrial Revolution through the current day. Over that time, there have been profound changes in knowledge about the science of climate change and its probable impacts, as well as in knowledge about alternatives such as solar energy. Thus, it is helpful to consider not just the general issues of when legal systems impose liability for past actions, but also the more specific one of how changes in knowledge can create retroactive liability.

In all three sections, we will be particularly attentive to the distinction between liability based on negligence or some other form of fault versus strict liability. If we applied a negligence regime to climate change, we would need to ask whether, during periods of the past, a reasonable person would have taken steps to limit emissions given the known costs and benefits of doing so. Under strict liability, this would no longer be the question, although even strict liability regimes may have exceptions based on the scientific "state of the art." In the final section, we give fuller consideration to the specific problem of historic responsibility for climate change.

A caveat is needed at the outset. No individual legal system is simple and unambiguous, whether we are talking about international public law, EU or U.S. federal law, or the laws of individual European nations or American states. These regimes, in turn, are obviously only a part of the global legal tapestry. Thus, any analysis necessarily shortchanges both the nuances and complexities of individual legal systems and the diversity of legal regimes. Nevertheless, it is illuminating to examine key features of prominent legal systems to provide a benchmark for analysis of climate change responsibility. As one would expect given the varied nature of the legal regimes in question, the lesson is not free of all ambiguity. Nevertheless, to the extent we give credence to these analogies, we seem to be led to embrace responsibility for some but not all historic emissions.

II Liability for Environmental Harm in International Law

International law is the obvious starting point in considering historic responsibility for a global problem such as climate change. It is important to keep in mind that the international regime, unlike national systems, does not have a legislature that can create binding legal rules, an executive that can enforce rules by force, or courts whose judgments must be obeyed on penalty of jail or fines. As a result, international law is heavily reliant on voluntary acceptance by the nations subject to its rules. No international tribunal can force the United States, the European Union, or China to pay damages for past emissions if they are unwilling to do so. On the other hand, a judgment from an international tribunal may give rise to a sense of obligation to comply or to give rise to pressure to comply from other nations.

A clarification about terminology is worthwhile at the outset. There is a technical distinction between duty and responsibility in international law. For instance, states may have a duty under international law not to harm ambassadors of other nations. The doctrine of state responsibility relates to a somewhat different question: what kinds of actions within its territory must a state answer for? For instance, is it responsible if a mob attacks a foreign ambassador, or if it becomes aware that a foreign intelligence service operating in its territory plans to harm the ambassador? As we will see, the distinction between state duties and state responsibility is blurred in international environmental law, which often speaks of a state's obligations to prevent certain types of actions by itself or within its territory or control.

International law requires states to limit activity that would be harmful to other states. In the *Trail Smelter* decision of 1941, an arbitration tribunal held Canada responsible for damage in the United States due to a smelter located near the Canada–U.S. border. The instructions given to the tribunal included the following direction:

> The Tribunal should endeavor to adjust the conflicting interests by some "just solution" which would allow the continuance of the operation of the Trail Smelter but under such restrictions and limitations as would, as far as foreseeable, prevent damage in the United States, and as would enable indemnity to be obtained, if in spite of such restrictions and limitations, damage should occur in the future in the United States.
>
> *(McCaffrey, 2006; Weiss, 2007: 257–58)*

In what has stood as an accepted principle of international law, the tribunal held that

> under the principles of international law, as well as of the law of the United States, no State has the right to use or permit the use of its territory in such a manner as to cause injury by fumes in or on the territory of another or the properties or persons thereof, when the case is of serious consequence and the injury is established by clear and convincing evidence.
>
> *(Weiss, 2007: 257–58)*

The *Trail Smelter* decision has been treated as one of "the foundation stones on which all international law is constructed"; although this mythic status may be exaggerated, the decision continues to be cited as "the *locus classicus* of international environmental law" (Bratspies and Miller, 2006). For instance, the 1992 Rio Declaration stated that states have "the responsibility to ensure that activities within their jurisdiction or control do not cause damage to the environment in other States or of areas beyond the limits of national jurisdiction." Similarly, a 2001 statement by the International Law Commission provides that states "shall take all appropriate measures to prevent significant transboundary harm or at any event to minimize the risk thereof." An alternative formulation by the International Law Association indicates that "States are in their legitimate activities under an obligation to prevent, abate, and control transboundary pollution to such an extent that no substantial injury is caused in the territory of another State," and to "endeavor to reduce existing transboundary pollution [that is already below that level] ... to the lowest level that may be reached by measures practicable and reasonable under the circumstances" (Weiss, 2007: 282–304).

Not surprisingly, there has been considerable debate about how these principles might apply in the context of climate change. Clearly, any effort to recover damages for climate change would encounter serious obstacles. Putting aside the problems of obtaining jurisdiction over an emitting state and of enforcing any resulting judgment (Koivurova, 2007), proof of liability would pose challenges in terms of establishing a causal link between specific emissions and particular damages (Adler, 2007). The barrier would be particularly high if proof by clear and convincing evidence is required, as stated in the *Trail Smelter* ruling (Faure and Peeters, 2011; Lord, 2012).

More relevant to the present discussion, claimants would face the problem of establishing that emitters violated a legal standard of conduct. Statements of the *Trail Smelter* rule tend not to be very specific on this score. Although climate change as a whole clearly meets the "substantial harm" threshold, it may be more difficult to assign responsibility for significant harm to any specific set of emissions from an emitter. Apart from claims based on the *Trail Smelter* principle, claims might also be brought for violations of treaty obligations undertaken under the United Nations Framework Convention on Climate Change (UNFCCC), although these are very vague, or under the Kyoto Protocol (but only for countries that ratified the protocol) (Faure and Nollkaemper, 2007).

A key question is whether it would be necessary to show that the emitting state was at fault. International law does not have a general rule requiring fault as a basis for liability if an action is attributable to a state and violates the state's international obligations. But it may be relevant because some international obligations require only reasonable efforts:

> In international law, liability does not depend on fault and is established on the basis of attribution and breach alone. Thus, a plaintiff would "only" have to show that an obligation of the Kyoto Protocol was breached. The situation is different when the primary obligation that is breached provides for a requirement of fault. Such is the case for the customary obligation that states should prevent transboundary damage. If the state has been diligent in regulating and controlling the harmful activity, yet transboundary damage still occurs, recourse against private actors is the only option left. That avenue depends on domestic law.
>
> *(Faure and Nollkaemper, 2007: 145–46)*

The question, then, is whether fault is required to establish a violation of the *Trail Smelter* rule (Weiss, 2007: 257–58). Although the issue is not free of dispute, Faure and Nollkaemper are probably right that some degree of fault is required.

International law is based largely on state consent to rules, and states have been reluctant to embrace strict liability for harm (Faure and Nollkaemper, 2007: 146). Treaty efforts aimed at imposing liability on states or even liability on private entities for transboundary pollution have received scant support, with the notable exceptions of measures dealing with oil spills. Developing states have tended to support such treaties, while developed countries have generally been resistant (Sachs, 2008). "States are reluctant to accept liability standards for environmental harm for obvious financial and political reasons, particularly in those states which might potentially be liable themselves for causing harm" (Tinker, 1992: 164).

This history suggests that developing countries will face practical barriers in attempting to hold developed countries responsible for all historical emissions, regardless of the time or the importance of the activities in question. There is a good argument that nations generally have recognized they have responsibility for causing environmental harm. But it is less clear that liability should be strict, in the sense of either ignoring the emitter's knowledge of the risks of climate change or the countervailing benefits of the activities leading to the emissions.

III Liability for Hazardous Waste and Other Environmental Harms

In contrast to climate change, many liability issues involve conduct by the defendant that is nearly simultaneous with injury to the plaintiff. Modern societies have learned, however, that hazardous wastes can pose very long term risks. Hazardous waste may leak into soil and groundwater long after the date of disposal. For this reason, hazardous waste presents issues analogous to climate change in terms of historic responsibility. We will begin by discussing the response to this problem in the United States before considering the European Union's approach. It should be noted at the outset that these regimes may impose strict liability in the sense that it is not necessary to show that the toxic release was due to carelessness on the part of the defendant. They do not, however, impose absolute liability because defendants may still be able to avoid liability under some circumstances.

A Liability under the U.S. Superfund Law

With the increased use of chemicals in the modern era, disposal of hazardous waste has become an increasing problem. Modern societies have imposed rules governing future disposal of waste. But they have also had to grapple with the question of who should pay for cleanup of past sites and for their environmental harms. As we will see, the United States has adopted a particularly rigorous liability scheme.

1 Introduction to CERCLA

CERCLA is the common acronym for the Comprehensive Environmental Response, Compensation, and Liability Act. The statute was passed in response to the dramatic discoveries in a number of locations, particularly the Love Canal site in New York, of dangerous levels of toxic substances in soil and groundwater, which stemmed from leaks at hazardous waste disposal sites. Congress swiftly took action to authorize government cleanups and to make the responsible parties liable for the costs (Salzman and Thompson, 2014: 247–48).

CERCLA imposes liability for cleanup and restitution costs on those persons who are responsible for the presence of the chemicals, without regard to fault or when their actions took place. Section 106 authorizes the attorney general to seek injunctive relief where an actual or threatened release poses an "imminent and substantial endangerment" to the public health or welfare or the environment. Alternatively, the president may issue administrative orders directing responsible parties to take protective action. Section 107 of CERCLA provides that generators and transporters of hazardous substances, as well as owners and operators of treatment, storage, or disposal (TSD) facilities, are liable for all costs of removal or remedial action incurred by the federal or state government "not inconsistent with" national standards for cleanup issued by the president; any other "necessary" response costs incurred by any other person, "consistent with" the plan; and (c) damages to publicly owned "natural resources" resulting from release of hazardous substances. Section 107(b) contains narrow exceptions for releases caused solely by acts of God or war, or by acts or omissions of certain "third parties." Thus, although liability is strict in the sense that no fault on the defendant's part is required, it is not absolute because it does excuse defendants when the release is clearly outside any possible control on their part.

Although exceptions from liability exist, they are very narrow. Consider, for instance, someone who acquires land with no knowledge that it was ever used for waste disposal. Under Section 101(40), such an "innocent owner" can avoid liability only if a series of other requirements are met. For instance, the person must have made "all appropriate inquiries into the previous ownership and uses of the facility in accordance with generally accepted good commercial and customary standards and practices" (Section 101(40)(B)). In addition, the person must use appropriate care to stop any continuing release and to limit human exposure to any previously released chemicals (Section 101(40)(D)); must cooperate fully with any cleanup activity (Section 101(40)(E)); and must have no direct or indirect business affiliation with any of the waste generators or past operators of the waste facility

(Section 101(40)(H)). In short, to be exempted from liability, the current owners must be not only free from fault but completely spotless.

Section 107(a) defines in general terms the costs and damages recoverable under CERCLA. It mentions "costs of removal or remedial action" incurred by the federal or state government, "other necessary costs of response" incurred by any other person, and "damages for injury to, destruction of, or loss of natural resources, including the reasonable costs of assessing such injury, destruction, or loss."

Under § 107(f), CERCLA liability for injury to natural resources is owed to the U.S. government; to a state government for resources "within the State or belonging to, managed by, controlled by, or appertaining to such State"; and to any Indian tribe in specified situations. Authority to sue is given to the president or the "authorized representative of any State," who "shall act on behalf of the public as trustee of such natural resources." Sums recovered shall be used by the trustee only to "restore, replace, or acquire the equivalent of" the natural resources injured, destroyed or lost. "Natural resources" are defined by § 101(16) to mean land, fish, wildlife, biota, air, water, groundwater, and other such resources belonging to, or managed or held in trust by, the United States, a state or local government, a foreign government, or an Indian tribe.

2 *Scope of Liability*

Liability under CERCLA is sweeping. For instance, it can potentially include an owner who bought the property after waste had been buried and began leaking, but before leakage was discovered (Salzman and Thompson, 2014: 253). A sense of the scope of liability can be seen in the treatment of waste generators. The courts have held that a waste generator can be held liable if the following four conditions are met: (1) the generator's hazardous substances were, at some point in the past, shipped to the facility; (2) the generator's hazardous substances, or hazardous substances *like* those of the generator, were present at the site at the time of a later leak; (3) there was a release or threatened release of *any* hazardous substance at the site; and (4) the release or threatened release resulted in response costs. Note that the plaintiff does not have to show that hazardous substances traceable to each defendant were released at the site. Requiring specific proof of causation would be at odds with the express language of the statute and would frustrate the intent of Congress because of the technological infeasibility of tracing a given generator's substances at the site (*United States v. Monsanto*, 858 F.2d 160 (4th Cir. 1988)).

CERCLA does offer defenses, but they are quite limited. As mentioned earlier, defendants can avoid liability by showing that release of waste was due to acts of God, war, or an independent third party. The third party must not have been in any direct or indirect relationship with the defendant. A *de micromis* exemption applies to defendants who contributed very small amounts of waste (less than about four hundred liters or one hundred kilos). Lenders are also generally exempt even if they ultimately own the site through foreclosure, and there are exemptions for landowners who did not dispose of waste and used due diligence to discover any possible waste issues before acquiring the property (Salzman and Thompson, 2014: 255–60). The narrow scope of these exemptions highlights the breadth of liability under the statute.

3 Strict Liability and Retroactivity

Some commentators have questioned the fairness of CERCLA:

> The biggest concern raised by the retroactive liability standards and broad categories of PRPs, though, is that of fairness. . . . It is worth considering whether there is a principled reason to hold many of the PRPs liable? Can they all fairly be thought of as polluters? And is it fair to hold them liable for actions taken years ago that were completely legal at the time? Where is the fairness, for example, in holding a chemical company liable today for shipping its waste back in 1950, in full compliance with the law, to an operating landfill that was improperly shut down in 1970? The traditional explanation for these liability provisions has been that owners, operators, generators, and transporters benefited from the low costs associated with waste disposal practices in the past and should bear the cleanup costs when they arise. You reap what you sow.
>
> *(Salzman and Thompson, 2014: 254–45)*

Despite these possible concerns, the courts have had no qualms about retroactive application of CERCLA to waste disposed before the statute's passage. In *United States v. Northeastern Pharmaceutical & Chemical Co., Inc.*, 810 F.2d 726 (8th Cir. 1986), the court relied on the language of the statute to supply a basis for retroactivity. As the Court explained:

> Although CERCLA does not expressly provide for retroactivity, it is manifestly clear that Congress intended CERCLA to have retroactive effect." The language used in the key liability provision, CERCLA § 107, 42 U.S.C. § 9607, refers to actions and conditions in the past tense: "any person who at the time of disposal of any hazardous substances owned or operated," CERCLA § 107(a)(2), 42 U.S.C. § 9607(a)(2), "any person who ... arranged with a transporter for transport for disposal," CERCLA § 107(a)(3), 42 U.S.C. § 9607(a)(3), and "any person who ... accepted any hazardous substances for transport to ... sites selected by such person," CERCLA § 107(a)(4), 42 U.S.C. § 9607(a)(4). *See, e.g., United States v. Conservation Chemical Co.*, 619 F.Supp. 162, 220 (W.D.Mo.1985); *United States v. Shell Oil Co.*, 605 F.Supp. at 1069–73; *United States v. South Carolina Recycling & Disposal, Inc.*, 20 Env't Rep. Cases (BNA) 1753, 1760–62 (D.S.C.1984); *United States v. A & F Materials Co.*, 578 F.Supp. at 1259; *United States v. Price*, 577 F.Supp. 1103, 1111–12 (D.N.J.1983); *Ohio ex rel. Brown v. Georgeoff*, 562 F.Supp. 1300, 1312 (N.D.Ohio 1983); *United States v. Outboard Marine Corp.*, 556

F.Supp. 54, 57 (N.D.Ill.1982); *United States v. Reilly Tar & Chemical Corp.,* 546 F.Supp. 1100, 1113–14 (D.Minn.1982); *see generally Developments in the Law-Toxic Waste Litigation,* 99 Harv.L.Rev. 1498 (1986) (*Developments*).

Further, the statutory scheme itself is overwhelmingly remedial and retroactive. CERCLA authorizes the EPA to force responsible parties to clean up inactive or abandoned hazardous substance sites, CERCLA § 106, 42 U.S.C. § 9606, and authorizes federal, state and local governments and private parties to clean up such sites and then seek recovery of their response costs from responsible parties, CERCLA §§ 104, 107, 42 U.S.C. §§ 9604, 9607. In order to be effective, CERCLA must reach past conduct. CERCLA's backward-looking focus is confirmed by the legislative history. *See generally* H.R.Rep. No. 1016, 96th Cong., 2d Sess., *reprinted in* 1980 U.S. Code Cong. & Ad. News 6119 (CERCLA House Report). Congress intended CERCLA "to initiate and establish a comprehensive response and financing mechanism to abate and control the vast problems associated with abandoned and inactive hazardous waste disposal sites."

(Id. *at 22, 1980 U.S. Code Cong. & Ad. News at 6125).*

United States v. Northeastern Pharm. & Chem. Co., Inc., 810 F.2d 726, 732–33 (8th Cir. 1986)

The defendants also argued that "retroactive" application of CERCLA, to create liability for conduct that occurred prior to the act's passage, would violate due process. The court rejected the argument, saying that a presumption of constitutionality attaches to "legislative Acts adjusting the burdens and benefits of economic life," and that such laws are constitutional unless the legislature acted in an arbitrary and irrational way: "Cleaning up inactive and abandoned hazardous waste disposal sites is a legitimate legislative purpose, and Congress acted in a rational manner in imposing liability for the cost of cleaning up such sites upon those parties who created and profited from the sites and upon the chemical industry as a whole." The court reaffirmed this ruling in 2001, rejecting arguments that an intervening Supreme Court case changed the applicable law (*United States v. Dico, Inc.,* 266 F.3d 864 (8th Cir. 2001); see also Whalin, 1999).

It would not be hard to make an analogy between CERCLA and historic carbon emissions, which (like leaking barrels of chemicals) can in effect poison the environment decades later. Thus, CERCLA provides a precedent of sorts for imposing strict liability for historic emissions. On the other hand, disposing of hazardous waste is an unusual activity, and waste generators and disposers were at least on notice that many of the chemicals were dangerous. Humans have produced carbon dioxide emissions, however, since they discovered fire, and societies did not take seriously even acute pollution risks until the final third of the twentieth century, let alone the less apparent potential for climate change. Thus, the analogy may be flawed, at least if we are concerned about whether entities were on notice about the need to exercise care.

B *Liability for Environmental Harm in the European Union*

Some EU member states have long-standing statutes dealing with environmental harm. For example, Germany has "a strict liability regime for damage caused to water and soil as well as a selected list of sites," and the United Kingdom "imposes strict liability for designated nature-protection sites" (Weisbach, 2012: 555). Since 2004, moreover, EU member states have been subject to an EU directive on environmental liability. It embodies the "polluter pays principle" in terms of remedying environmental damage (Bergkamp and Goldsmith, 2013). This section explains the features of the directive that may have lessons for thinking about historic carbon emissions. The directive was clearly inspired in part by CERCLA, but also differs from the American liability statute in important ways.

1 *Introduction to the EU Directive*

The 2004 directive addresses damages to natural habitats and species, which early drafts of the directive described as "biodiversity damage." It also addresses soil pollution that creates a significant risk of harm to human health. This coverage is analogous to that of CERCLA. In addition, the directive covers water pollution (Brans, 2013: 31–38).

Damages include "(1) the cost of restoring the injured natural resources and services to baseline; (2) the cost of restoration that compensates for the interim loss of resources and services that occur from the time of the incident until recovery of such resources and services to baseline; and (3) the reasonable cost of assessing damages" (Brans, 2013: 39). There is no cap on the amount of liability. In addition, in cases of imminent threat, firms have an obligation to take necessary preventive measures (Brans 2013: 46–57).

Entities subject to the directive are termed "operators," which is a narrower group than the potentially responsible parties covered by CERCLA. The key is that the operator is the entity with control over an activity. Operators in some activity sectors that were considered potentially dangerous to the environment are subject to strict liability, while other firms are liable only on the basis of fault. The standards that trigger strict liability are listed in an annex to the directive and are quite broad:

> The triggers that bring an activity under Annex III [strict liability] are diverse and include the use of particular substances or technology, size, emissions, etc. These activities, which are described in broad terms in the Annex, include, for instance, "manufacture, use, storage, processing, filling, release into the environment and onsite transport of dangerous substances" ... There is no threshold and thus any use or storage, even in negligible quantities, of any dangerous substance, even relatively safe products, would

appear to render the activity in which they are used subject to the ELD [the Directive]. To take another example, use and storage of plant protection or biocidal products causes activities to be subject to the ELD's strict liability regime.

(Bergkamp and van Bergeijk, 2013a: 63)

At least in the context of hazardous waste, liability for operators may be as broad as liability under CERCLA. Consider the first case decided by the European Court of Justice (ECJ) under the directive, where there was no direct evidence linking the defendant and the site:

In the suit brought by the Italian Economy Ministry against the refinery companies, the European Court of Justice ascertained liability based on the fact that a chemical used by the refineries was found present on the contaminated site. The litigation resulted from the contamination of the Augusta Harbor off the coast of Sicily. Several chemical companies have operated their facilities adjacent to the harbor for decades. Italian authorities eventually ordered cleanup of the harbor, which had collected more than two meters of sediment pollution. Upon establishing the link of causation, the refinery and all current landowners were ordered to pay for damages and necessary measures to contain the contamination.

(Tabatabai, 2012: 680)

There is no limit on the interval between the incident and the time when authorities issue a restoration order giving rise to liability (Bergkamp and van Bergeijk, 2013a: 70–71). However, liability for damage payments is limited to a thirty-year period. In addition, the Directive exempts "diffuse pollution," meaning situations in which harm is produced by multiple operators and it is impossible "to establish a causal link between the damage and the activities of individual operators" (Bergkamp and van Bergeijk, 2013b: 84). Thus, the directive clearly would not cover climate change.

Overall, the directive "arguably contributed to the realization of the polluter pays and preventive principles (EC Treaty, Article 174(2)), and to ensuring the decontamination and restoration of the environment, including across Member States' borders" (Faure and De Smedt, 2013: 308). However, because member state implementation of the directive was very slow, there still have been relatively few cases of its use in practice.

2 *Liability Limitations*

Most strikingly, the directive is not retroactive. It does not apply to damage caused by any event before April 30, 2007. It also excludes events that occurred after that date but were caused by a specific activity that was finished earlier. It is not clear exactly what constitutes an "activity" – for example, whether putting material in storage is the relevant activity or whether instead the continuing act of storage is itself an activity (Bergkamp and van Bergeijk, 2013b: 86). This would matter if chemicals were placed into storage more than thirty years ago but the leakage continued past the thirty-year mark.

The directive also allows member states to adopt two additional defenses that limit the scope of strict liability. First, member states may exempt from strict liability an emission or event authorized by certain types of permits. Second, member states can exempt certain types of actions including

> an emission or activity or any manner of using a product in the course of an activity which the operator demonstrates was not considered likely to cause environmental damage according to the state of scientific and technical knowledge at the time when the emission was released or the activity took place.
>
> *(Bergkamp and van Bergeijk, 2013b: 91)*

Even when the two defenses apply, however, operators are still liable for negligence, provided that that the failure to exercise due care caused the environmental harm in question. Fewer than half of the EU's members adopted both defenses (including the United Kingdom, Spain, and Italy), whereas others (including Germany) decided not to allow either one. France adopted only the state-of-the-art defense but not the permit defense (Bergkamp and van Bergeijk, 2013b: 93).

The Environmental Liability Directive provides a weaker basis than CERCLA for arguments in favor of historic responsibility for carbon emissions. The directive's lack of retroactivity is an obvious stumbling block. The halfhearted rejection of the state-of-the-art defense also suggests that historic responsibility would at best be limited to the time after carbon emissions were known to pose a risk of dangerous climate change. At most, if the directive were taken as a guide to the appropriate approach, historic responsibility for carbon would extend back only thirty years.

IV Products Liability

Like hazardous waste, defective products can cause long-delayed harm. Some products, such as industrial equipment, are durable and remain in use for many years. Other products, such as industrial chemicals or pharmaceuticals, can cause latent damages that blossom into health problems after many years. Thus, the law of products liability can pose some similar questions to climate change about the scope of responsibility for acts committed well in the past. Products liability law is a relatively modern development. To put it in perspective, we begin by considering the general rules that govern liability for harmful activity, and then turn to the special rules for defective products in the United States and the European Union.

A The Default Requirement of Fault

The United States is a common law jurisdiction. Accordingly, tort law is largely the product of judicial decisions. In general, tort law imposes liability on the basis of negligence. That is, defendants are liable when their failure to take reasonable precautions results in harm. The key to establishing negligence is to demonstrate that a reasonable person would have taken greater precautions. Compliance with industry custom is strong evidence of reasonable care, but the ultimate question is whether the foreseeable risks would have justified greater precautions whether or not the industry as a whole took such precautions. Since foreseeability is the key, the question must be assessed on the basis of the risks that the defendant could have reasonably foreseen and the precautions available at the time, not with the benefit of hindsight. Often the degree of the reasonable precaution is determined by reference to a rough cost–benefit analysis, taking into account the cost of precaution, the probability of harm, and the degree of harm (Abraham, 2012: 58–81; Diamond, 2008: 154–92). For instance, according to one highly influential source, "primary factors to consider in ascertaining whether the person's conduct lacks reasonable care are the foreseeable likelihood that the person's conduct will result in harm, the foreseeable severity of any harm that may ensue, and the burden of precautions to eliminate or reduce the risk of harm" (American Law Institute 2010: 62).

Apart from the area of products liability discussed later, it is unusual for U.S. courts to impose liability under the common law without negligence. However, courts have recognized a small area of tort liability for "ultrahazardous" or "abnormally dangerous" activities that cannot be made sufficiently safe even with the use of reasonable care. As the use of the word "abnormally" suggests, one key factor is the unusualness of the activity given the time and place. For example, courts have applied strict liability for damages caused by use of explosives or collapse of privately owned water reservoirs, but not for transportation of toxic substances in railroad tank cars (Abraham, 2012: 197–204). A few American courts have rejected strict liability for hazardous activities altogether (Diamond, 2008: 507). Although formulations of the test vary, a recent authoritative summary of the law states that an activity is abnormally dangerous if "(1) the activity creates a foreseeable and highly significant risk of physical harm even when reasonable care is exercised by all actors; and (2) the activity is not one of common usage" (American Law Institute, 2010: 276). Carbon emissions do not generally qualify for strict liability under this standard, since burning fossil fuels is (unfortunately) a matter of common usage.

European states similarly emphasize fault as the background norm for liability (Lord et al., 2012). English law is similar to U.S. law (Goldberg and Lord, 2012), which is not a coincidence since U.S. tort law was rooted in the English common law. The terminology in continental Europe is different: wrongful acts are called delicts rather than torts. In Germany, damages are provided under § 823 of the Civil Code, with negligence defined under § 276 as the failure to exercise reasonable care. Reasonable care can be established by showing that an expert confirmed the absence of a significant risk. Under this approach, a "person has not acted with reasonable care when he/she has not undertaken necessary and reasonable action to avoid a foreseeable danger" (Koch, Luhrs, and Verheyen, 2012: 408).

Unlike in the English and American common law systems, liability in continental Europe is based on legislatively enacted codes rather than judicial decisions (Wagner, 2008). Liability is based on fault, which is defined in objective terms. For instance, in French law

> the objective conception of fault also prevails in the French doctrine of *faute civile*, from which the so-called *élement moral* has been deliberately purged in favour of an appreciation *in abstracto*. The benchmark for the general legal duty to take care is the conduct of a fictitious reference person, the bon père de famille, not the actual person whose behavior is evaluated.
>
> *(Wagner, 2008: 1024)*

Under this standard, however, "dangers that would have been invisible to a reasonable person in the situation of the tortfeasor before the accident occurred do not require precautionary measures, and measures of safety which were not (yet) available in the historical situation must likewise be ignored" (Wagner, 2008: 1026).

As in American law, strict liability is also imposed in some circumstances. In most civil law countries, strict liability is relatively minor, but French law is unusual, recognizing strict liability for accidents caused by objects under the control of the defendants. Such custodial strict liability would virtually swallow all of the law of delicts, but the French courts have drawn a distinction between active and passive contributions to damages:

> French jurists, interpreting what constitutes the "act of a thing" (*le fait de la chose*) under article 1384, also resort to an artificial notion. They invoke the distinction between the active and passive role: if the thing played only a passive role in causing the plaintiff's damage, it would not be considered the cause of the damage. For instance, if a vehicle leaves the road and crashes into a wall, it would be absurd to a French jurist to say that the custodian of the wall is responsible for the damage to the vehicle or its driver under article 1384 ... On the other hand, if a thing is in an abnormal condition (a car with defective brakes) or an abnormal position (parked in the middle of the road), it plays an active role even though inert at the time of the accident. It is then considered the cause of the damage.
>
> *(Palmer, 1988: 1326)*

Arguably, "this distinction reintroduces the concept of fault under the guise of causation because the abnormal behavior of the thing often coincides with evidence of the carelessness of the *gardie*." (Palmer, 1988: 1326). According to another commentator, "proof of passivity bears a close resemblance to proof of due care in using the thing in a normal manner" (Tomlinson, 1988: 1345).

With the arguable exception of France, then, strict liability has been a deviation from the default rule of negligence in the United States and Europe. The dominant rule has been liability based on fault, with foreseeability of risk as a key element. Applying these concepts in the context of climate change would require that responsibility for past emissions be conditioned on the foreseeability of climate damage relative to the benefits of using fossil fuels. As we will see, however, the scope of responsibility might be greater, however, if the analogy is to modern laws governing liability for harm due to manufactured products.

B Products Liability in the United States

In the United States, the major exception to the negligence requirement under the common law is liability for personal injury from defective products. This section considers the scope of this exception and how products liability applies to risks that were unforeseeable when a product was produced.

1 Scope of Liability

Until the 1960s, liability for harm caused by defective products was generally governed by negligence law in the United States. Courts began to turn increasingly toward strict liability for defective products that posed unreasonable danger to consumers or the public. Courts were concerned because of the decreased ability of consumers to protect themselves by investigating products in a more technologically advanced world, and also by the difficulty of establishing the seller's negligence in many situations. Strict liability was appealing as a way to provide greater incentives for the seller to exercise care and to allow remaining product risks to be spread among consumers through price increases. Thus, it became clear that sellers or manufacturers of defective products would be strictly liable for harm to consumers (Abraham, 2012: 216–28).

The move to strict products liability made it crucial to determine when a product should be considered defective. A swimming pool is not defective merely because it is possible to drown in a pool of water. When a product causes harm because of an error in the manufacturing process, it is relatively easy to determine that the product is defective, because it deviates from the intended product design, as can be confirmed by comparison with the defendant's normal production. But it is more difficult to determine whether the design itself is defective. Courts have tended to decide this question on the basis of whether the product lives up to the safety expectations of the consumer or by comparing the risks posed by the design with the design's utility, generally compared with other possible designs. A related theory of liability is that warnings are needed to make the product reasonably safe to use (Abraham, 2012: 227–36).

In general, even if a product is defective, liability requires showing that the defendant's product caused the plaintiff's injury. In exceptional circumstances, some American states have accepted a theory of "market share" liability. This theory applies when many manufacturers have sold identical defective products – for example, a pharmaceutical product that turns out to cause cancer. Because of the long time between use of the product and development of cancer, it may no longer be possible to determine which manufacturer's product caused a specific plaintiff's harm. Under these circumstances, some states have essentially imposed liability on the manufacturers as a group, divided up according to their relevant market shares. This approach to liability does show the willingness of courts to be creative when needed to prevent injustice, but it is important to note that market share liability addresses the difficulty of proving a causal link between the defendant's conduct and the plaintiff's harm. It does not address the initial question of whether the defendant's past conduct was of the type to trigger liability, because the whole argument is premised on the assumption that all of the defendant's products were defective.

2 The State-of-the-Art Defense

Some products may remain in use long after their initial sale, and products may also cause delayed harm after use, as in the case of chemicals that cause cancers that appear many years later. These time lags open the possibility that new risks will be identified or new design improvements invented in the meantime. Thus, a fundamental question is whether the defectiveness of a product design or a warning should be determined solely on the basis of the state of the art when the product was produced or sold, or whether scientific and technological progress between that time and the injury are also relevant. In short, even if a product's designs or warnings would be considered defective if the product were manufactured today, should the seller or manufacturer be able to defend its conduct on the basis of the earlier state of the art at the time of sale?

U.S. courts are not in complete agreement about the answer to this question (Diamond, 2008: 548–57). For instance, in *Robinson v. Brandtjen & Kluge, Inc.*, 500 F.3d 691 (8th Cir. 2007), the operator of a printing press brought a product liability claim against the manufacturer of a printing press after her hand was smashed between the printing plates. The question was whether a detachable guard should have been provided for situations when the operator was manually feeding paper into the printer. The printer had initially been sold in 1940 to a newspaper and was resold to the plaintiff's employer some fifty years later. The court ruled in favor of the manufacturer:

> We come to this case in 2007, but we must consider whether there is sufficient evidence to support a finding that the B & K press was defective when it was sold more than sixty-five years ago in 1940. We find instructive the analysis of the Fourth Circuit, in similar circumstances ... That court observed that it is inappropriate to superimpose contemporary standards of safety on an earlier era: "In short, a product can only be defective if it is imperfect when measured against a standard existing at the time of sale or against reasonable consumer expectations held at the time of sale."

Some states have adopted this view by statute, though in some cases the statutes have been so poorly drafted as to be ineffective. For example, Kentucky law provides that "in any product liability action, it shall be presumed, until rebutted by a preponderance of the evidence to the contrary, that the product was not defective if the design, methods of manufacture, and testing conformed to the generally recognized and prevailing standards or the state of the art in existence at the time the design was prepared, and the product was manufactured" (*Sexton v. Bell Helmets, Inc.*, 926 F.2d 331 (4th Cir. 1991)). This provision turned out to be meaningless, since the presumption was automatically overcome whenever a plaintiff could offer convincing proof that a defect existed on the basis of the current state of knowledge (ibid.).

In contrast, the state-of-the-art defense was rejected by the court in *Sternhagen v. Dow Company* (935 P.2d 1139 (Mont. 1997). The defendant manufactured a herbicide known as 2,4-D. The plaintiff's husband was exposed to the herbicide in 1948–50 and was diagnosed with cancer some thirty years later, allegedly as a result of the exposure. The court was asked to decide whether the manufacturer was conclusively presumed to know the dangers inherent in the product or whether state-of-the-art evidence is admissible to establish "whether the manufacturer knew or through the exercise of reasonable human foresight should have known of the danger." The court held the evidence inadmissible:

> To recognize the state-of-the-art defense now would inject negligence principles into strict liability law and thereby sever Montana's strict products liability law from the core principles for which it was adopted – maximum protection for consumers against dangerous defects in manufactured products with the focus on the condition of the product, and not on the manufacturer's conduct or knowledge.
>
> (Sternhagen v. Dow *at 1144)*

One difficulty in interpreting these cases is that the "term 'state of the art' has been variously defined to mean that the product design conforms to industry custom, that it reflects the safest, and most advanced technology developed and in commercial use, or that it reflects technology at the cutting edge of scientific knowledge" (American Law Institute, 1997: 81; for more detailed analysis, see Robb, 1982). The judicial decisions can be divided into several categories:

- "A few states take the position that conformance with the best available technology in actual use is an absolute defense to design liability."
- Others "take the position that evidence that a risk was beyond the scope of scientific knowability at the time of manufacture is inadmissible."
- Courts often hold that conformity with the best existing design is not an absolute defense if an "alternative design proffered by the plaintiff could have been practically adopted by the manufacturer at the time of sale" (American Law Institute, 1997: 81–84).

The law seems to remain unsettled, but the majority view seems to require a pragmatic assessment of what reasonable expectations can be placed on the manufacturer taking into account contemporary knowledge and practices at the time of manufacture. Thus, the majority trend is to allow consideration of the state of the art (Diamond, 2008: 557).

C Products Liability in the European Union

Prior to 1985, products liability was governed by the laws of EU member states. Consequently, it varied from place to place. In Italy, fault of the manufacturer could be assumed from the occurrence of the accident. In Germany, the burden of proof to establish negligence was reversed so that the manufacturer was required to establish the exercise of due care (Pasa and Benacchio, 2005: 101–10). In France, although fault-based liability was the default rule, it had limited application because of the potential for contractual liability or liability imposed on custodians of things (which was sometimes extended to render a manufacturer a continuing custodian for dangerous products) (Whittaker, 2005: 50–57). As discussed later, these preexisting rules were modified after the 1985 EU directive (Whittaker, 2005; Fairgrieve, 2005).

1 Scope of Liability under the 1985 Directive

The 1985 European Directive on Products Liability established the principles of European law on the subject. Article 6 provides that a "product is defective when it does not provide the safety which a person is entitled to expect, taking into account all circumstances" (Stapleton, 2002: 1231–32). This determination is to be made "taking all circumstances into account, including (a) the presentation of the product; (b) the use to which it could reasonably be expected that the product would be put; and (c) the time when the product was put into circulation."

However, a "product shall not be considered defective for the sole reason that a better product is subsequently put into circulation" (Whittaker, 2005: 481). Damages recoverable under the directive include personal injury and property damage (Whittaker, 2005: 502).

A key question is the extent to which the directive controls the laws of member states. Article 13 provides that member states may not adopt new laws derogating from the convention, whether in favor of the plaintiff or the defendant. It does, however, allow traditional rules of liability to remain in place. Specifically, the directive "shall not affect any rights which an injured person may have according to the rule of the law of contractual or noncontractual liability or a special liability system existing at the moment when this Directive is notified" (Whittaker, 2005: 437). The European Court of Justice (ECJ) issued three decisions in 2002 that attempted to clarify the application of the directive. They make it clear that the laws passed in member states specifically to implement the directive may not deviate from its requirements. They leave in doubt, however, the extent to which liability broader than the directive can be established under other legal theories (Whittaker, 2005: 440–45). Indeed, French consumers may well continue to rely on preexisting legal theories that are more favorable than the directive itself (Taylor, 2005).

Although the directive is phrased in terms of consumer expectations of safety, courts have "noticed that consumer expectations are not of much guidance in a courtroom where there simply are no consumer expectations or the existing expectations are unrealistically high or lag behind the level of safety achieved by the latest technology" (Lenze, 2005: 209). It is sometimes said that consumers are entitled to expect that a product is "state of the art," but "the understanding of this concept varies considerably, ranging from 'the cutting edge technology available at the time' to nothing more than 'good industry practice'" (Lenze, 2005: 110).

2 *Foreseeability of Harm*

Section 7 is known as the development risk provisions, and it allows a manufacturer to escape liability if it can show that "the state of scientific and technical knowledge at the time when it put the product into circulation was not such as to enable the existence of the defect to be recognized." Countries are not required to adopt this defense, however, which was inserted into the directive as an option at the demand of the Thatcher government in the United Kingdom (Stapleton, 2002: 1231–32). Only Finland and Luxembourg have omitted the state-of-the-art defense entirely, but Germany and Spain have done so with respect to pharmaceuticals (Mildred, 2005: 168).

Independent commissions unsuccessfully argued that the United Kingdom should not implement the § 7 defense (Mildred 2005: 167). In implementing the directive, the United Kingdom rephrased it as exonerating a defendant if "the state of scientific and technical knowledge at the relevant time was not such that a producer of products of the same description as the product in question might be expected to have discovered the defect if it had existed in products while they were under his control." Note that this language goes beyond whether the general risk was known into the question of its detectability in particular products. The ECJ rejected the claim that this language unduly broadened the defense (Whittaker, 2005: 495–99).

The major dispute over the defense is whether it requires "absolute undiscoverability" or "undiscoverability by reasonable means." Court decisions on the subject are sparse. Courts have split on whether risks that were known to exist but were not detectable or avoidable are covered by the defense. The ECJ has said, however, that the state of knowledge is not limited to what a particular producer actually knew or had reasonable means to learn, but rather on the objective state of scientific and technological knowledge at the time (Mildred, 2005: 177). Such knowledge must be accessible, meaning at least that unpublished work does not count (Mildred, 2005: 177–91).

The dominant approaches to products liability in the United States and the European Union might suggest by analogy some limits on climate responsibility. In particular, they suggest that climate responsibility might be cut off prior to the time that the risks of climate change were understood. After that time, however, something closer to strict liability might apply, at least to the extent that reduced use of fossil fuels would have been feasible.

V Implications for Climate Change Responsibility

In considering what lessons comparable legal regimes may have in the climate context, we might begin by asking what historic responsibility for carbon emissions would look like under various legal regimes. Under a negligence regime, it would be necessary to show that historic emitters were at fault: that they should have foreseen the harms caused by carbon emissions and taken steps to prevent those harms. In considering emissions from a past year, we would have to ask questions about the degree of harm that a reasonable person would have foreseen and the feasibility of preventing that harm, taking into account the costs and benefits for doing so. Such a standard might make it difficult to establish responsibility further in the past, when knowledge of climate change might not have made clear the extent of potential harm, the need for economic growth may have been strong, and alternatives to fossil fuels may not have been available. Even under a fault approach, however, at least some responsibility for past emissions would exist. For instance, by 1990, the international regime had taken notice of climate change and alternatives such as wind power, solar energy, nuclear energy, and improved energy efficiency were recognized. To the extent that a reasonable person, taking into account the costs and benefits of carbon emissions, would have taken steps to reduce emissions, failure to do so would lead to responsibility for those incremental past emissions.

In contrast, under a regime of strict liability, it would be easier to establish responsibility. It would not be necessary to show that nations were at fault for failing to control emissions earlier. But strict liability is not the same as absolute liability. A regime of strict liability might allow a "state-of-the-art" defense if, during the time period in question, the scientific evidence did not yet establish the dangers of climate change or alternative feasible energy technologies did not exist.

As we have seen, there is some precedent for each of these approaches in analogous situations. Liability based on fault seems to be the default rule. The findings in this article are consistent at a general level with an earlier, less in-depth examination of existing legal practice:

> From this brief survey, we can see that while there is some precedent for using a strict liability standard in contexts similar to climate change, it is quite limited. Even in environmental contexts, most countries require fault most of the time.
>
> *(Weisbach, 2012: 556)*

Nevertheless, there are some important qualifications to be made. Stricter forms of liability are not infrequent, such as CERCLA in the United States and the Environmental Liability Directive in the European Union. Defenses based on the state of the art are common but clearly are not universal. And courts have also made some provision for strict liability, exemplified by liability for abnormally dangerous activities in the United States and custodial liability in France. There is a continuum of fault, from basic negligence, on one end, to failure to research risks as deeply as humanly possible, on the other, so it is not necessarily enough to simply distinguish negligence from strict liability.

International law has been particularly cautious about embracing strict liability for environmental harm. The reluctance of developed countries to agree to such liability, even in situations where the financial stakes are much lower compared with climate change, may suggest that it is unrealistic to expect them to do so in the context of carbon emissions.

Analysis of EU and U.S. liability rules in analogous areas suggests the need for separate consideration of two issues. The first issue is whether liability should attach for risks that were not foreseeable at the time. Despite the rejection of the state-of-the-art defense in some jurisdictions, the clear tendency seems to be to require consideration of foreseeability. Either under strict liability or negligence, existing liability regimes suggest that responsibility probably should not attach for emissions prior to the time when a reasonable person could have known that emissions posed a threat of serious harm. Such knowledge was surely present by the late twentieth century, and a reasonable person might well have appreciated the existence of a possibly serious risk somewhat earlier.

The second issue is whether the risks should be balanced against the benefits of the conduct. In the hazardous waste setting, such balancing has not been accepted, while balancing is seemingly inevitable in determining whether a risk renders a product "defective." The hazardous waste cases might be explained as an outgrowth of the category of abnormally dangerous activities, or they might reflect the difficulty of reconstructing what safety measures were used in waste disposal decades earlier. Although the argument for a balancing approach may be a bit stronger on the basis of these areas of the law, they do not provide a clear message about whether the utility of past carbon emissions should weigh in the balance.

Even basic negligence, however, would leave some room for taking historic emissions into account. Despite arguments to the contrary about the difficulty of determining negligence in the context of climate change (Weisbach, 2012: 552), use of the risk utility approach would make this conceptually straightforward (if perhaps more difficult in practice). Under the risk utility approach, we would ask the level of carbon emissions that would have been justified taking into account the risks and benefits of an activity. This is equivalent to determining the social cost of carbon (based on the information available at various times in the past), and then asking what the level of the activity would have been if the actor had had to pay the social cost of carbon. The difference would be the amount of carbon negligently emitted into the atmosphere. Such calculations would reduce the amount of carbon that would be considered each country's historic responsibility, but at least some past carbon would still be covered.

These various liability schemes can provide only some helpful guidelines for thinking about responsibility in the unique context of climate change. Furthermore, as we have seen, those indications are not unambiguous. They probably provide somewhat more support for the developed country position for the period before the gravity of climate risks was known, but their message is more ambivalent about whether a balancing test is needed to determine the level of responsibility after that point.

Thus, it would be something of an uphill battle to argue on the basis of existing legal norms that developed countries should be responsible for all the carbon remaining in the atmosphere that they have ever emitted. On the other hand, there is nothing in analogous rules about liability that would support a complete exemption from responsibility for historic emissions. The judgments societies make in contexts involving current harm from past conduct suggest the need to balance the claims of victims to compensation against the claims of emitters to fair notice of the consequences of their actions. More succinctly, analogous legal regimes suggest strongly that the right answer is somewhere between full responsibility for all historic emissions and complete exoneration for those emitters.

The two intermediate regimes are strict liability, with a state-of-the-art defense, and negligence liability. Under one, responsibility would begin at the point when the existence of significant climate risks began to be within the state of the art scientifically. The other would begin at the point when a reasonable emitter would be aware of the existence of such risks.

The difference between the two intermediate approaches would reflect the lag time between awareness of the dangers of climate change in the scientific community and awareness in the larger world. It is difficult to attach clear dates to these standards. Hunter and Salzman provide a helpful timeline of the relevant events (Hunter and Salzman, 2007). Although there were earlier hints of concerns in the nineteenth century, the first serious warnings from individual scientists were issued in the mid-1950s. By the late 1970s and 1980s, international conferences were being held, and by 1990 the Intergovernmental Panel on Climate Change issued a formal warning of the danger of climate change. So the possibility of climate risks presumably was within the state of the art at some point between around 1960 and 1990. By 1992, the UNFCCC had called for international action to reduce emissions, with a focus on developed countries. In terms of the negligence standard, Hunter and Salzman suggest that it became unreasonable to expand emissions without taking into account climate change around the turn of the twenty-first century (Hunter and Salzman, 2007: 1773). This seems to be at the upper edge of reasonable estimates. More than a decade earlier, the framework convention had articulated an international consensus on the danger of climate change and the need to control emissions. This should have placed a reasonable person on notice of the problem. As the scientific evidence firmed up and became more widely known, the reasonable person would have taken increasing precautions.

As discussed earlier, negligence is generally the default rule of liability. Under a negligence standard, emitters would be responsible for supporting some kind of action to deal with emissions by 1995, give or take a decade. Emitters would then be responsible for excess emissions after that point, meaning emissions that a reasonable person would have prevented (Faure and Nollkaemper, 2007: 145, 151).

The other intermediate possibility would be represented by the "state-of-the-art" approach. Under that approach, the cutoff date for liability would be somewhat earlier, and emitters would be responsible for all emissions after that date, not just the ones they should have reasonably prevented.

Although the possibility of applying the "state-of-the-art" approach cannot be excluded, applying it to climate change would be harsher than applying it in the context of products liability where it originated. By triggering liability based on the state of the art, rather than on what a reasonable person would know, the "state-of-the-art" approach normally gives an industry an incentive to keep abreast of scientific developments relating to its products and even to sponsor such research. But applying such an approach to a problem like climate change, which does not relate to an emitter's own products but rather to its energy sources, seems to set an unrealistic expectation.

Thus, the products liability approach does not seem to present a compelling analogy to climate change. This leaves the negligence approach as the most plausible possibility, situated between the extremes of no responsibility for historic emissions or of responsibility stretching back to the beginning of the Industrial Revolution.

References

Abraham, K. (2012). *The Forms and Functions of Tort Law*, 4th ed. St. Paul, MN: Foundation Press.

Adler, M. (2007). Commentaries: Corrective Justice and Liability for Global Warming. *University of Pennsylvania Law Review*, **155**, 1859–67.

American Law Institute (1997). *Restatement of the Law of Torts: Products Liability*, 3rd ed. St. Paul, MN: American Law Institute Publishers.

American Law Institute (2010). *A Concise Restatement of Torts*, 2nd ed. St. Paul, MN: American Law Institute Publishers.

Bergkamp, L., and Goldsmith, B. (2013). *The EU Environmental Liability Directive: A Commentary*. Croydon, UK: CPI Group.

Bergkamp, L., and van Bergeijk, A. (2013a). Scope of the ELD Regime. In *The EU Environmental Liability Directive: A Commentary*, ed. L. Bergkamp and B. Goldsmith. Croydon, UK: CPI Group, pp. 51–79.

Bergkamp, L., and van Bergeijk, A. (2013b). Exceptions and Defences. In *The EU Environmental Liability Directive: A Commentary*, ed. L. Bergkamp and B. Goldsmith. Croydon, UK: CPI Group, pp. 80–94.

Brans, E. (2013). Fundamentals of Liability for Environmental Harm under the ELD. In *The EU Environmental Liability Directive: A Commentary*, ed. L. Bergkamp and B. Goldsmith. Croydon, UK: CPI Group, pp. 31–50.

Bratspies, R., and Miller, R. (2006). *Transboundary Harm in International Law: Lessons from the Trail Smelter Arbitration*. Cambridge: Cambridge University Press.

Diamond, J. (2008). *Cases and Materials on Torts*, 2nd ed. St. Paul, MN: WestAcademic.

Fairgrieve, D. (2005). *Product Liability in Comparative Perspective*. Cambridge: Cambridge University Press.

Farber, D. (2008). The Case for Climate Compensation: Justice for Climate Change Victims in a Complex World. *Utah Law Review*, **2008**(2), 377–413.

Faure, M., and Nollkaemper, A. (2007). International Liability as an Instrument to Prevent and Compensate for Climate Change. *Stanford Environmental Law Review* **23A**, 123–79.

Faure, M., and Peeters, M. (2011). *Climate Change Liability*. Cheltenham, UK: Edgar Elgar.

Faure, M., and De Smedt, K. (2013). The ELD's Effects in Practice. In *The EU Environmental Liability Directive: A Commentary*, ed. L. Bergkamp and B. Goldsmith. Croydon, UK: CPI Group, pp. 219–234.

Goldberg, S., and Lord, R. (2012). England. In *Climate Change Liability: Transnational Law and Practice*, ed. R. Lord et al. Cambridge: Cambridge University Press, pp. 445–88.

Hunter, D., and Salzman, J. (2007). Negligence in the Air: The Duty of Care in Climate Change Litigation. *University of Pennsylvania Law Review*, **155**, 1741–94.

Koch, H. J., Luhrs, M., and Verheyen, R. (2012). Germany. In *Climate Change Liability: Transnational Law and Practice*, ed. R. Lord et al. Cambridge: Cambridge University Press, pp. 376–416.

Koivurova, T. (2007). International Legal Avenues to Address the Plight of Victims of Climate Change: Problems and Prospects. *Journal of Environmental Law and Litigation*, **22**, 267–99.

Lenze, S. (2005). German Product Liability Law: Between European Directives, American Restatements, and Common Sense. In *Product Liability in Comparative Perspective*, ed. D. Fairgrieve. Cambridge: Cambridge University Press, pp. 100–25.

Lord, R. et al. (2012). *Climate Change Liability: Transnational Law and Practice*. Cambridge: Cambridge University Press.

McCaffrey, S. (2006). Of Paradoxes, Precedents, and Progeny: The *Trail Smelter* Arbitration 65 Years Later. In *Transboundary Harm in International Law: Lessons from the Trail Smelter Arbitration*, ed. R. Bratspies and R. Miller. Cambridge: Cambridge University Press, pp. 34–45.

Mildred, M. (2005). The Development Risk Defense. In *Product Liability in Comparative Perspective*, ed. D. Fairgrieve. Cambridge: Cambridge University Press, pp. 167–191.

Nottage, L. (2004). *Product Safety and Liability in Japan from Minimata to Mad Cows*. London: Routledge Curzon.

Palmer, V. (1988). A General Theory of the Inner Structure of Strict Liability: Common Law, Civil Law, and Comparative Law. *Tulane Law Review*, **62**, 1303–53.

Pasa, B. and Benacchio, G. (2005). *The Harmonization of Civil and Commercial Law in Europe*. New York: CEU Press.

Robb, G. (1982). A Practical Approach to Use of State of the Art Evidence in Strict Products Liability Cases. *Northwestern University Law Review*, **77**, 1–33.

Robinson v. Brandtjen & Kluge, Inc., 500 F.3d 691 (8th Cir. 2007).

Sachs, N. (2008). Beyond the Liability Wall: Strengthening Tort Remedies in International Environmental Law. *UCLA Law Review*, **55**, 837–904.

Salzman, J. and Thompson, B. H. (2014). *Environmental Law and Policy: Concepts and Insights*, 4th ed. St. Paul, MN: Foundation Press.

Sexton v. Bell Helmets, Inc., 926 F.2d 331 (4th Cir. 1991).

Stapleton, J. (2002). Bugs in Anglo-American Products Liability. *South Carolina Law Review*, **53**, 1225–61.

Sternhagen v. Dow Company, 935 P.2d 1139 (Mont. 1997).

Tabatabai, M. (2012). Comparing U.S. and EU Hazardous Waste Liability Frameworks: How the EU Liability Directive Competes with CERCLA. *Houston Journal of International Law*, 34, 653–84.

Taylor, S. (2005). Harmonisation or Divergence? A Comparison of French and English Product Liability Rule. In *Product Liability in Comparative Perspective*, ed. D. Fairgrieve. Cambridge: Cambridge University Press, pp. 221–43.

Tinker, C. (1992). Strict Liability of States for Environmental Harm: An Emerging Principle of International Law. *Transnational Law*, **3**, 155–66.

Tomlinson, E. (1988). Tort Liability in France for the Act of Things: A Study of Judicial Lawmaking. *Louisiana Law Review*, **48**, 1299–1360.

United States v. Dico, Inc., 266 F.3d 864 (8th Cir. 2001).

United States v. Monsanto Co., 858 F.2d 160 (4th Cir. 1988).

United States v. Northeastern Pharmaceutical & Chemical Co., Inc., 810 F.2d 726 (8th Cir. 1986).

Wagner, G. (2008). Comparative Tort Law. In *The Oxford Handbook of Comparative Law*, ed. M. Reiman and R. Zimmerman. Oxford: Oxford University Press, pp. 1003–42.

Weisbach, D. (2012). Negligence, Strict Liability, and Responsibility for Climate Change. *Iowa Law Review*, **97**, 521–65.

Weiss, E. B. et al. (2007). *International Environmental Law and Policy*, 2nd ed. New York, NY: Walters Kluwer Publishers.

Whalin, D. (1999). Is there still pre-1980 CERCLA liability after *Eastern Enterprises*?. *Environmental Lawyer*, **5**, 701–73.

Whittaker, S. (2005). *Liability for Products: English Law, French Law, and European Harmonization*. Oxford: Oxford University Press.

5

Asking Beneficiaries to Pay for Past Pollution

ANJA KARNEIN

Climate change is no longer a future scenario. Already today, its effects are palpable. Several extreme weather events in the past few years across the globe, such as heat waves, cold spells, floods, mudslides, droughts, tornadoes, hurricanes, and typhoons, have been attributed to climate change, leading to the injury, illness, and death of millions of people. The rising sea levels and the increased frequency of small tidal waves have forced thousands to leave their homes for good, such as the inhabitants of the Carteret Islands in the South Pacific[1], arguably creating a new category of displaced persons: climate change refugees. Extreme weather events have also brought about increased human tensions. The armed conflict that led to the humanitarian crisis in Darfur in 2003, for instance, has been attributed to failing rains and creeping desertification as effects of climate change (Welzer, 2008).

Apart from events of current and acute disaster, a less stable climate leads to more unpredictable and extreme weather, causing everyone who is left unprepared highly vulnerable to future catastrophe. Thus, beyond giving rise to the need for help in emergencies after the fact, climate change requires helping the most vulnerable countries adapt[2] to severe atmospheric conditions. In response to this, Article 4.4 of the United Nations Framework Convention on Climate Change (UNFCCC) states that "the developed country Parties and other developed Parties included in Annex II shall . . . assist the developing country Parties that are particularly vulnerable to the adverse effects of climate change in meeting costs of adaptation to those adverse effects." There seems to be

[1] This claim is not uncontroversial (and is thus no different from all other claims linking particular weather events to climate change). Some argue that rather than the sea levels rising, the problem for many islands is that they are sinking. This, in turn, may be part of their natural evolution.

[2] According to the UNFCCC, adaptation means "adjustment in natural or human systems in response to actual or expected climatic stimuli or their effects, which moderates harm or exploits beneficial opportunities." See: http://unfccc.int/essential_background/glossary/items/3666.php.

widespread agreement that this kind of aid needs to be given and that the wealthy, developed countries should provide it. What remains subject to debate, however, is on what grounds such assistance should be made available.

Interestingly, the convention treats the obligation to help developing countries adapt to the detrimental effects of past pollution akin to the obligation to help developing countries adapt to the detrimental effects of current pollution. This is noteworthy given that these cases are frequently treated as distinct in the literature (Caney, 2010; Meyer and Roser, 2011; Page, 2012). The argument is that while current polluters are clearly morally responsible[3] and are around to pay their bill, the same cannot be said about a number of past polluters. This is because they – arguably, as a result of their ignorance regarding the detrimental effects of CO_2 – cannot be held morally accountable and are no longer alive to compensate for the damage they caused. Thus, supporting adaption to the consequences of past pollution may have to occur for different reasons than those invoked to ground the obligation to help countries adapt to current pollution (the detrimental consequences of which may not appear until sometime in the future).[4]

In response to this difficulty, some have argued that once the original polluters are no longer available, we should resort to asking those who benefited from their pollution to pay toward compensating the harm thereby created (see, for instance, Gosseries, 2004: 43–45; Shue, 2010: 105; Page, 2012). Asking beneficiaries of past pollution to pay appears promising indeed: it identifies duty holders without needing to assign any blame (Page, 2012: 306–07, 317), and it is not as arbitrary as many consider applying the so-called Ability-to-Pay Principle (APP) to be, which requires the wealthy to pay just because they are wealthy (Page, 2012: 307, 311, 316). Moreover, it is intuitively appealing: why would somebody be entitled to gains on account of an action (in this case: pollution) that is causing others to suffer undeserved harm? At the same time, however, asking

[3] I am assuming that moral responsibility requires having voluntarily engaged in an action that one could have reasonably expected to be dangerous or detrimental to someone else, either by design or as a foreseeable side effect. Suppose someone has a habit of throwing little cotton balls into the air. There would be no reason for him or her to expect anybody to be physically harmed by this action. If, because of some completely unforeseeable event, one of these cotton balls ultimately hurt people (perhaps in the far future), the original thrower would clearly be causally responsible but not morally. I am assuming that the case of past emissions is similar – past polluters arguably did not or could not have reasonably been expected to know that their actions would have adverse consequences (Singer, 2002: 34; Miller, 2008: 129–30).

[4] This assumes that there is indeed a difference between past and present collective agents (generations) and that we do not treat such agents as parts of transhistorical collective entities (e.g., countries). But even if we view countries as transhistorical collective entities, the amount of guilt accumulated before, say 1990, is presumably less than afterward since certain excuses that applied before (ignorance, for instance) do not do so after. The only difference is that, when viewed this way, polluters and beneficiaries are one and the same entity and the distinct consequences of having benefited become less easy to disentangle. As it is my goal to analyze the problems and merits of the Beneficiary Pays Principle (BPP), I will treat past and current generations as distinct, even if that may seem artificial in some ways. I will address this concern explicitly in section IV.

beneficiaries to pay is also rather perplexing: why exactly should the mere fact of having benefited from past pollution generate obligations of some kind?

The question this chapter addresses is whether it is convincing to hold those who benefited from past harmful emissions responsible for providing aid or compensation toward current efforts to adapt "solely because the disadvantages and benefits share common origins" (Page, 2012: 313). I argue that having benefited alone cannot provide the necessary grounds for establishing such duties. I show that additional factors have to be present for benefits to generate obligations. Once these additional factors have been identified, however, it is no longer clear that having benefited remains relevant to the argument. This is not meant to excuse the rich and powerful, who tend to be the beneficiaries of past pollution, from having to fulfill their responsibilities toward those who need help in their efforts to adapt to climate change; rather, my argument is intended to make sure the reasons they are held to account are good ones. And asking them to pay just or primarily because they are better off than they would have been had pollution not occurred is not a valid reason, or so I claim.

The argument proceeds in four steps. The first step discusses three "pure" cases of the Beneficiary Pays Principle (BPP) in order to illustrate that benefits, by themselves and without additional arguments, do not generate duties. The second step presents cases that may seem as if benefiting generates duties to aid the victims. It shows that duties in such circumstances arise only if additional factors are brought into play and thus do not follow from having benefited alone. The third step surveys cases in which benefiting from the past harmful acts of others may seem to generate duties to compensate the victims – especially if the original perpetrators of the harm are no longer alive. I illustrate why any duties arising in such instances also crucially hinge on factors other than benefiting. The fourth step addresses two possible objections to the individualistic framework offered here, namely, (a) that we should think of the relevant agents as transhistorical collective units and (b) that particular duties arise not because individual agents have benefited but because we are involved in a global economic structure that systematically benefits some while it disadvantages others. I show that (a) may be an adequate description from some perspectives but does nothing to help solve the question at issue in this chapter, namely, whether the BPP has any merit as an independent principle in the climate change debate. With regard to (b) I claim that this may be true of part of the problem connected to poorer countries' increased vulnerability to climate change. I also say, however, that this does not change anything about my contention that benefiting from an unjust global structure in and of itself – without taking into account additional factors – does not generate obligations.

I Three "Pure" Versions of the BPP

In the following, I will present three of what I call "pure" versions of the BPP. What makes them pure is that the beneficiary has no connection to the harm beyond benefiting from it.[5] What changes from case to case is merely the cause of the harm and with it, perhaps, the strength of the intuition that something needs to be done to help or compensate the victim.

1. Rain: It rains. As a consequence, the local ice-cream parlor owner is worse off than she would have been had the sun been shining while the local movie theater owner is better off than she would have been had the weather been fairer.[6]

This case probably bothers few. It would be difficult to argue that the beneficiary (here the movie theater owner) owes anything to the harmed party (here the ice-cream parlor owner) just because of the weather. That is at least partly because raining seems to be a "natural" event with which both parties would be expected to have to reckon. Moreover, the mere circumstance that someone is comparatively better off after a certain (in this case, natural) event than she would have been had some other (again, in this case, natural) event occurred does not seem to be particularly interesting, at least not normatively speaking.

Would it make a difference if it turned out that not only was the victim harmed – that is, made relatively worse off than she would have been – but that, because of the rain, she has fallen below a minimal threshold of well-being? This further fact may certainly persuade some that help ought to be provided. But the question remains: by whom? The beneficiary is not an obvious choice: having benefited alone does not automatically render anyone well off absolutely speaking, nor guarantee that the beneficiary ultimately will be above the minimal threshold of well-being.

2. Accident: A geoengineer accidentally makes it rain, thereby making the local ice-cream parlor owner worse off than she would have been had the weather not been tampered with, but without intending or foreseeing this or the further consequence, namely, that the rain

[5] As already indicated in note 4, in all cases I discuss I am taking as given that the beneficiary is distinct from the perpetrator of the harm.

[6] Note that I am using comparative counterfactual notions of harm and benefit here (where someone is rendered worse or better off by some event or action than she *would* have been had the event or action not occurred). If I were to use historical comparative accounts (where someone is rendered worse off than she *was* before an event or action occurred) the case might not be as clear and thus not a particularly well-chosen one: if the ice-cream parlor owner sold even one ice cream that day, she would end up better off than she was before and thus, it could be argued, was not harmed at all. Were I to use normative notions (where someone is rendered worse or better off than she *should* have been), the same question I am asking would pose itself, namely, whether or why, even if everyone were to agree that the harmed party ought to be compensated, the person compensating for it should be the beneficiary as opposed to anyone else. But the point here is to provide a case in which someone was harmed while another was benefited – I leave it up to the reader to decide on her favorite notion of harm and benefit. For further reading on these various conceptions of harm see, for instance, Meyer 2015, section 3.1.

will make the local movie theater owner better off than she would otherwise have been. The geoengineer has left the scene and is nowhere to be found.

Obviously, this situation is unfortunate. What has changed from *Rain* to *Accident* is that (a) the source of the loss is no longer a mere event but an attributable action, and (b) that, although the loss to the ice-cream parlor owner may be exactly the same as in the previous example, the nature of this loss is different: it is now arguably a rights violation (even though the geoengineer is presumably not culpable).[7] So we might feel that the victim deserves compensation. But even if we think that, it would not be immediately clear, at least not without further argument, that it should be the beneficiary who has to pay, as opposed to other agents capable of helping.

3. Intentional Rights Violation: A geoengineer makes it rain so that the local ice-cream parlor owner ends up worse off than she would have been. She thereby (accidentally) renders the local movie theater owner better off than she would have been. The geoengineer has left the scene and is nowhere to be found.

Here we have an instance of intentional (culpable) wrongdoing by the geoengineer. Unlike in the previous two cases, you might think that the beneficiary should provide the compensation in the absence of the perpetrator, since she has not only benefited, but also benefited from a culpably wrongful act. Surely the intuition will be strongest in this case. But again, without any further information, there is nothing to distinguish this case from the previous cases: We have a victim who, possibly more obviously so than in the other two cases, deserves to be compensated.[8] But even if the agent responsible for providing this compensation is gone and we thus want someone to compensate in the perpetrator's stead, it is not clear, at least not barring further information, why this should be the beneficiary in particular.

Thus, in both cases, that of asking beneficiaries to provide aid and that of asking beneficiaries to provide compensation, we need to look for additional factors that may make it the case that benefits generate obligations. In the following I survey several initially promising arguments for why beneficiaries of an action in the past owe something to the victims of this action, either (in section II) as a matter of providing aid or (in section III) as a matter of compensation.

[7] Actions that are rights violations can be described as cases of wrongful harm. I am assuming that there are three kinds of cases: first, "harm" in its purely descriptive usage, which does not necessarily infringe on any rights. Here, the harm does not need to be compensated (e.g., harm that ensues to a person's having to listen to children constantly screaming at the top of their voices in the neighborhood). Second, nonculpable wrongful harm, which infringes on rights (e.g., harm that ensues from one person's damaging the property of another, even if accidentally); past pollution presumably falls into this category of cases. Third, culpable wrongful harm, when someone is morally to blame for her actions that intentionally infringed on another's rights.

[8] This again is debatable. You might just think that this is a case in which the loss should lie where it falls.

II Benefits and Providing Aid

Consider *Rain* again. In this case I claimed that, without further information, we have no idea whether this rain undermines or promotes distributive justice more generally. But suppose again that it emerges that the ice-cream store owner not only suffers relative losses but ultimately falls below a minimum threshold of well-being and there is thus a compelling reason that aid be provided by *someone*.[9] In such cases one may think that beneficiaries are an obvious place to start looking because their benefit stands in some connection to the harm. David Miller, for instance, thinks that connections of certain kinds matter and offers what he calls the Connection Theory.[10] This holds that, in cases when it is clear that aid needs to be provided and thus it is just a question of who should be responsible for providing it, we should check the links established by the various different forms of responsibility at our disposal. Miller discusses six in particular, namely, causal responsibility, moral responsibility, capacity to pay, community, outcome responsibility, and benefit.[11] We should then decide, on a case by case basis, which applies most or, in Miller's own terms, which establishes the strongest connection to the victim. "Any of these relations . . . may establish the kind of special link between A [the agent] and P [the victim] that enables us to single out A as the one who bears the responsibility for supplying the resources that will remedy P's condition" (Miller, 2001: 470).

Miller claims that there are usually separate moral considerations at play in each specific situation that will emphasize one connection over another. If asking the person morally responsible for a harm to help the victim contributes to restoring justice between her and the victim or forces her to acknowledge the wrong she has committed, then this connection will be fairly strong (Miller, 2001: 470). Causal responsibility is the only case in which Miller thinks there are no separate moral considerations that would advise reliance on it. However, he contends that the reason we frequently do (and should) hold responsible the agent who "merely" caused a state of affairs to occur is that we ultimately need to hold someone responsible and, in the absence of any stronger connections, we have to resort even to rather weak kinds of

[9] Note that for this reason to exist it does not matter whether the cause of the victim's plight is a mere event (e.g., rain) or an attributable action (for which, however, the perpetrator, as far as she is culpable, is not available to provide aid).

[10] The situation he has in mind is the "current plight of Iraqi children who are malnourished and lack access to proper medical care" (Miller, 2001: 453). However, his principal concern is with situations, such as this one, when it is clear that aid needs to be provided and the question is only who has the responsibility to do so.

[11] Miller added outcome responsibility and benefit to his original list in *National Responsibility and Global Justice* (Oxford University Press, 2007), 100–01. In his revision, outcome responsibility is second – after moral and before causal responsibility (100–01) and benefit is fourth (102–03).

connection (Miller, 2001: 470–71).[12] In this way, Miller adequately describes why we will often be ready to hold those responsible for compensation whose connection to the harmful act is relatively feeble and when we also lack any other moral considerations to lay responsibility on them.

According to Miller, "benefit itself can serve as a ground of remedial responsibility: being a beneficiary of the action or policy that harmed P establishes a special connection with P of a kind that stands independently alongside the other forms of connections that make up this list, and that may in certain cases provide a decisive reason for A to remedy the harm that has befallen P" (Miller, 2008: 103). Interestingly, Miller claims that a person or agent A has benefited when "resources that would otherwise have gone to P have been allotted to A" (Miller, 2008: 102). That, however, describes only a very special case of benefiting. Rather than pointing to a situation of someone benefiting from an action that also happens to harm another, he has in mind someone (A) possessing something that rightfully belongs to another (P). This sounds more like a case of unjust enrichment and carries with it an array of difficult questions. Among them certainly is how A came by her new possessions: by way of a just transfer (e.g., acquiring a bicycle in a bicycle shop that later turns out to have been stolen), by luck (someone finds an unlocked bicycle in the street that appears to have been abandoned but turns out to have been stolen), or by more suspect means (someone buys a bicycle at the flea market or from the back of a bus that turns out to have been stolen). If, for instance, she bought a stolen bicycle in what appears to be a legitimate shop for bicycles, it is not obvious that the original owner of the bicycle has a claim on her – as opposed to a claim on the thief or, in her absence, on society as a whole.[13] While this is a highly intriguing and important discussion, it turns on questions directly related to moral responsibility[14] and thus does not speak to the questions at issue here, namely, whether a

[12] Simon Caney seems to be assuming something similar when he develops his "hybrid solution," a combination of the Polluter Pays Principle and the Ability-to-Pay Principle, which ensures that all the costs to the victims are covered (Caney, 2010).

[13] Immanuel Kant famously denies this, but for primarily pragmatic reasons related to ensuring the certainty of the law (*Rechtssicherheit*) (Kant 1797 (1996), 6:301–6:03). This is a powerful concern given the practical impossibility of establishing just title to most things we care about without it. So although I cannot consider this problem in more detail here, my short response in such a case would be that, quite apart from any pragmatic concerns, once the stolen good has been passed on in this (legitimate) way, taking it away again would appear to make the beneficiary the (second) victim. Thus, once such a transfer occurs, there are two interests to keep in mind: that of the original owner and that of the new one. Otherwise the legitimate ownership of the original owner would, in many cases, have to be called into question as well.

[14] This is certainly true of the way Miller thinks of it. More generally, however, this claim is controversial; some would like unjust enrichment to refer to cases of "mistaken payment" in which nobody has to be morally responsible for the benefit to create a duty to pay it back (Birks, 2003). But in such cases someone is at least causally to blame for making a mistake and I think it is not clear why it should not be *that* person who should have to bear the costs (although I would agree that it would be a sign of good character for the person benefited by the mistake to pay back the benefit she received).

beneficiary of an event or action that causes another harm is sufficiently "connected" merely by having benefited to acquire responsibilities to provide aid.

So the question is whether a less "responsibility-loaded" account of benefiting than the one Miller appears to have in mind can establish the kind of connection that would generate duties. There are two reasons this seems doubtful. The first is pragmatic and the second normative. The pragmatic concern has to do with the purpose of asking beneficiaries to pay, which, in the case of Miller's argument, is to provide relief to the victims. If this is the purpose, however, then beneficiaries are a poor choice since beneficiaries of harmful actions are usually held responsible for paying only in proportion to how much they benefited and not more (Butt, 2007: 142; Gosseries, 2004: 47).[15] This is to ensure that someone who only benefited insignificantly does not eventually have to pay a large sum of aid or compensation. But if benefiting is not necessarily understood as an instance in which "resources that would otherwise have gone to P have been allotted to A" (Miller, 2008: 102), the responsibility for providing aid acquired by the benefit is disconnected from the harm it is supposed to address and, with it, from the assurance that the amount someone benefits coincides with the amount another loses. Thus, whatever can be asked of beneficiaries will only ever help to relieve the victim's plight by chance, that is, if it happens to be the case that someone benefited to the same extent that someone else was harmed. In any particular case, however, an agent's benefit may very well be insufficient to remedy the victims' situation (Butt, 2007: 141–42).

Asking the beneficiary to pay would not only be a poor choice pragmatically speaking. By singling out beneficiaries as having a special connection to the victim, one is assuming something that still needs to be proven, namely, that benefiting constitutes not only some sort of connection besides others, but one that, more than causal responsibility, has separate moral considerations speaking for it. The beneficiary would have to stand out somehow from all possible connections agents may have to the event or the victim. But there is nothing particularly distinctive about the connection of having benefited: the only thing this signifies is that an event or action had different consequences for two distinct parties. After all, the harm did not occur *because* of the benefits. And, second, presumably, whenever someone benefits, there are a number of other agents who have relationships that "establish the kind of special link between A [the agent] and P [the victim] that enables us to single out A as the one who bears the responsibility for supplying the resources that will remedy P's condition" (Miller, 2001: 470).

[15] To be clear: this is not a necessary feature of the BPP. One could, for instance, also ask beneficiaries to pay for the entire damage regardless of how much they benefited. But not only is limited liability (limited by the extent of the benefit) endorsed by many supporters of the BPP, it is so for a reason: asking more would seem unfair to the beneficiaries since it is the benefit that supposedly is generating the liability and nothing else.

To begin with, she may have family, a neighbor whom she once helped out, close friends, the community more generally, etc. And all these people might also have benefited, just from other events, actions, or omissions. Thus, it is really not so obvious why we should single out the beneficiaries among these various connections, at least not without additional information.

What would such additional information have to look like? We might, for instance, think that of all these relevantly similar connections (and with no one causally responsible), those agents should pay who are most able to remedy the situation. Not only would they get the job done: they would also be the ones on whom the effects of doing so would be least severe. These considerations would provide independent moral grounds for favoring this connection over the others. So if it turned out that the beneficiaries were the ones most able to help, we might be led to ask them. Note, however, that what would be doing the work here is the ability to pay and not having benefited. It would not, for instance, matter why the beneficiaries were most able to pay, that is, whether they acquired this ability by benefiting or quite independently.

In conclusion, it appears that applying the Connection Theory to the understanding of benefit offered here provides no reasons for choosing beneficiaries of particular events or actions as being responsible for providing aid to those who might end up being harmed by these events or actions – unless, that is, it happens also to be the case that these beneficiaries are most able to pay. But then nothing really hangs on having benefited. Applied to the case of past pollution this implies that, barring any further information, the mere fact that some benefited does not in and of itself render them liable to provide aid to those suffering the consequences (more, that is, than anyone else would be so liable).

This having been said, benefiting from past pollution may not appear to be like the case just portrayed: past pollution is not merely an "event" such as rain, from which some happen to benefit while others are harmed. Rather, past pollution is due to attributable actions, namely, those of polluters. Thus, let me now turn to cases that raise the issue of whether benefiting can ever give rise to duties of compensation.

III Benefits and Compensation

The idea that benefits from past pollution can create duties of compensation to the victims of past pollution may arise in two ways.[16] First, one may think that it is

[16] Compensation is usually owed when someone has been harmed or has suffered a loss at the hands of someone else. There are various ways of being harmed or of suffering a loss at the hands of someone else. As the three "pure" cases I presented in section I showed (*Rain, Accident,* and *Intentional Rights Violation*), only some involve the harmful action also being wrongful (accidental rights violation), and of these again only some are also culpable (intentional rights violation). If, and this is what I am assuming, we do not think that only culpably wrongful harms need to be compensated, the following discussion will show that while the distinction

wrong to benefit from past injustice and that whoever does so acquires a duty to compensate the victims of such injustice. Second, one may come to hold that beneficiaries of past injustice ought to pay compensation because they are part of the reason that the harm was done in the first place.

1. *The Duty to Condemn an Injustice*. Consider the following case. Suppose a stranger walks into a bar, is in an especially good mood, and invites everyone to a round of free beer. Imagine that a day later these benefited guests learn that this stranger violently stole this money from his elderly landlady, who was badly wounded in the course of the crime. By the time this information surfaces, the culprit has disappeared and is nowhere to be found. Obviously he should be punished and owes his landlady some form of compensation for her harm as well as a sincere apology. In his absence, do the beneficiaries have a duty to compensate the landlady? Daniel Butt thinks so. He claims that one has a duty not to benefit from an unjust act (Butt, 2007: 143) and that "we make a conceptual error if we condemn a given action as unjust, but are not willing to reverse or mitigate its effects on the grounds that it has benefited us. The refusal undermines the condemnation" (ibid.: 143). Accepting a benefit of an injustice thus makes us violate our duty to condemn unjust actions. Interestingly, however, Butt goes on to say that "if no one else is willing or able to make up these losses [suffered by some on the basis of an unjust action], then the duty falls to those who are benefiting from the distortion in question" (ibid.: 143). This suggests that if the original culprit (the stranger, in our case) is available and can compensate in full, the beneficiaries (the patrons of the bar, in our case) may not have to compensate for anything (ibid.: 142), but that, in his absence, they do. That is strange: Why would an obligation for the beneficiaries arise just because nobody else is available? If benefiting from an unjust act is in itself morally wrong, as Butt insists it is, would it not have to create obligations on its own account and regardless of whether the original wrongdoers are still around or not?[17]

Moreover, there is a question of how this independent offense should be addressed. Here one could pick up on Butt's suggestion and have the beneficiary pay in proportion to the benefit she received. In our case, this would presumably mean asking the beer drinkers to pay back the price of their beers. Would this then not be a genuine instance in which we ask beneficiaries to pay? Probably not. This is because the main problem is that the beneficiaries committed their own separate offense, not that they benefited. They did not have to benefit in order to render

between nonwrongful and wrongful harm may change the way the necessity of compensation is intuitively felt, this distinction is irrelevant when it comes to determining whether benefits generate duties of compensation.

[17] Caney has asked the same question (Caney, 2006: 472–73).

themselves guilty. Had they merely applauded the injustice and bought the stranger a drink instead of being invited (i.e., provided, as opposed to reaping, a benefit) they would have made themselves at least as guilty, and perhaps even more so. Thus, the degree of their wrong does not correspond to the benefits received (or provided, as the case may be). So while we may think that individuals ought to give up benefits received in the course of committing the wrong of not condemning an unjust act, they may be asked to give up more because the benefits neither (a) were a necessary means by which the wrong was committed nor (b) in any way automatically correspond to the degree of wrongfulness involved.

Now let us assume, for the sake of argument, that we have just encountered a way in which benefits play at least some role in generating certain obligations, perhaps by means of their symbolism: regardless of what else the patrons of the bar owe to the landlady if they knowingly held on to the free beer, giving up the benefits derived from this beer may be thought to be a necessary part. If this is so, then the question is whether benefiting from past pollution is like knowingly accepting benefits from a past wrongful action and thereby acquiring a duty that may also involve having to divulge any benefits received. It is not. The crucial difference is that past pollution cannot so clearly be classified as an injustice that one has a duty to condemn. Emitting CO_2 is not akin to committing a crime. This is for two reasons (both are necessary; neither one is sufficient): to begin with, pollution only becomes problematic if everyone does it to extensive degrees and over long periods (Sinnott-Armstrong, 2010: 333–34). Second, while it is true that past polluters took more than their fair shares of a limited resource (the atmospheric capacity to absorb CO_2), we cannot conceive of them as having committed an injustice because, while they ended up being factually wrong, the evidence at the time rendered this mistake a reasonable one to make.[18] Thus, even if duties are generated by knowingly failing to condemn a wrongful act, this is not necessarily what happens when agents benefit from past pollution.[19]

2. *Being Implicated in a Wrongful Act.* We might expect benefits to create obligations if the beneficiaries of harmful actions are somehow actively involved

[18] As I say in the text, both conditions have to hold, that is, both that (a) the action in question is not wrongful as such and that (b) the evidence at the time did not make it unreasonable to think that it was innocent. The same could not be said about slavery, for instance, since here (a) does not apply: enslaving a person as such *is* wrongful.

[19] Here is where Edward Page's defense of the BPP is confusing. Although he admits that past emissions were neither blameworthy nor wrongful, his argument for asking beneficiaries to pay seems to rest – at least in part – on the argument that beneficiaries, by holding on to their benefits, fail to condemn a past injustice: "The BPP, grounded in the deeper principle that *agents should not profit from injustice*, requires that states disgorge a proportion of the accumulated 'climatic benefit' they have received so that it can be used to *remediate* the suffering with climate change accruing in other states that have not shared in the profits of industrialization to the same extent and who also now face the prospect of constraints on their use of the assimilative capacity of the atmosphere to prevent further climate change" (Page, 2012: 317, emphasis added AK).

in generating these harms. Axel Gosseries offers a fictional example of the triangular relationship for three island communities, the United States (which pollutes the atmosphere by producing goods), the European Union (which does not pollute but buys products from the United States), and Bangladesh (which is the victim of pollution). He claims that the European Union owes Bangladesh compensation in accordance with the benefits it received (Gosseries, 2004: 43). For Gosseries the principal reason for the European Union's obligation is that the European Union is a free rider on Bangladesh: it benefits from the pollution of the United States without having to bear the associated costs (i.e., to Bangladesh).

To begin with, nothing in this analysis hangs on whether the European Union actually benefits or not: Suppose the European Union did not benefit from importing products from the United States (because, for instance, Europeans had come to find these products so ethically offensive that nobody buys them and businesses are boycotted). This would not change our judgment about the European Union: as long as it is buying products from the United States, it is implicated in the wrongful harm caused to Bangladesh. This, I think, already suggests that it is not benefiting that is doing the work but rather the European Union's blameworthy complicity in injustice.

Gosseries argues that this case is essentially the same as the following one he offers (as a further step in his argument). Here the United States is held to be a past generation while the European Union signifies those currently living. As before, Gosseries suggests that the European Union owes Bangladesh compensation in accordance with the benefits it received because it is free riding on Bangladesh. If the United States is taken to be a past generation, however, the issue looks quite different. Then Bangladesh will have been harmed by past pollution while the European Union will be enjoying benefits that it did not ask for and cannot be expected to renounce. Although it continues to be true that Bangladesh deserves compensation, it is no longer clear that the European Union is involved in harming Bangladesh or has any other meaningful connection to the latter. This would be different only if it could be shown that a past U.S. generation polluted *in order to* benefit the European Union and that the European Union had an honest choice about whether to accept these benefits or not. If the European Union therefore could be held morally responsible (at least in part) for past pollution, the European Union itself would be implicated in the harmful action taking place and would, as Gosseries maintains that it does, incur compensatory duties toward Bangladesh. However, the European Union would not be a free rider in this instance, but would rather be implicated in the original harmful act.

In any case, past generations probably did not pollute because they thought their descendants would appreciate this particularly or that they would expect such benefits. Unless, therefore, one could make the case that past generations (mistakenly) understood it to be their duty to future persons to industrialize in the way they did, those currently living and benefiting from past pollution cannot be held responsible for their ancestor's actions. I think it is fair to assume that the latter emitted CO_2 in the first instance to improve their own situation and that the fact that some continue to benefit from these emissions today is a mere side effect. Thus, the BPP fails here as well. Either beneficiaries owe nothing, or, if they do owe something, this is due to their moral responsibility of some kind, either because they (a) committed a separate offense or (b) are implicated in a wrongful action. Benefiting from past pollution involves neither (a) nor (b).

IV Are Past Polluters and Current Beneficiaries Ever Really Distinct?

One criticism of the arguments presented thus far may be that it artificially draws a distinction between past polluters and present beneficiaries. Rather, we should think of transhistorical collective units as the relevant agents, such as nations (Miller, 2008: 128). If we think in those terms, the entire problem of accounting for the pollution of past generations seems to dissolve because original polluters will always also be current beneficiaries. But then nobody is saying anything about beneficiaries having to pay and my argument is directed only at those who do make such a claim.

This having been said, the case of transhistorical collective agents is not completely straightforward either. Although nations can be blamed for their past wrongdoing, this still does not change the fact that the climate change effected by past pollution is not the result of injustice. Even someone such as Miller, who argues that the relevant agents are nations and not individuals, therefore dismisses the notion of historical responsibility (Miller, 2008: 137). One might, of course, hold nations responsible simply for their nonwrongful past pollution (i.e., for their causal responsibility) according to the Polluter Pays Principle (PPP) and ask them to compensate the victims of this past pollution. This, however, has no relation to asking the beneficiary to pay. It is perhaps also worth mentioning that because we generally think that polluters are not like criminals whose punishment for their crimes is inalienable, we will only ever ask polluters to pay if they are able to do so. In that sense, the PPP can also only be applied in cases in which past polluters happen to be wealthy (irrespective of the source of this wealth).

There is a second reason why we might think that the tripartite relationship of A performing an action that harms B and benefits C is inadequate. One may object that the problem underlying the unequal distribution of undeserved harms and benefits of past pollution is a structural one: that the world economy is organized in such a way that the northern, industrialized, and wealthy nations are regularly benefited while the southern, developing, and poor nations are systematically disadvantaged. This is what makes it much harder for the latter to deal with the detrimental effects of climate change. Moreover, it is this unfair structure that obliges those who are regularly benefited to assist those who are systematically disadvantaged. This would be similar to the argument Thomas Pogge makes about why the wealthy nations are obliged, as a matter of compensation, to help the global poor. "Acceptance of such costs is not generous charity, but required compensation for the harms produced by unjust global institutional arrangements whose past and present imposition by the affluent countries brings great benefits to their citizens" (Pogge, 2005a: 741).

It is certainly true that the unfortunate circumstance that some nations are too poor to deal adequately with the detrimental effects of climate change is due to factors that extend beyond the pollution of individual actors sometime in the past. Furthermore, on this picture, the benefits that some enjoy today are never as "innocent" or "pure" as I have been assuming because, by benefiting from an unjust global structure, the beneficiaries are contributing to it. If this is the claim, however, I would respond exactly the way I have been arguing so far. I would contend that the reason this gives rise to obligations is that the beneficiaries are contributing to an unjust structure and not because they are benefiting. Whether anyone actually benefits or not seems to be irrelevant.[20] Depending on how one thinks the unjust structure works exactly, this would be similar to one of the two cases discussed earlier, namely, that agents, by benefiting, either compound an injustice (as in 1) or are directly implicated in it (as in 2). In neither case, however, does the mere fact that agents are benefiting generate duties of any particular kind.

Finally, there are at least two parts to the problem of climate change for especially vulnerable countries, only one of which can be captured by referring to unjust structures. The other part is related to the mere bad luck of being geographically located, namely, in the Southern Hemisphere, so that one is likely to be hit hardest by extreme weather phenomena. For that part of the problem the picture I have been painting certainly remains accurate (if we do not assume transhistorical collective units as agents): there is no necessary connection between the beneficiaries of the harmful actions (past pollution) and the harm that ensues. Therefore, without further information, it is not clear that those benefiting from past pollution should be the ones who ought to be made primarily responsible for

[20] Pogge would disagree with this. He argues that the fact that we profit "renders more stringent our obligation to compensate" (Pogge, 2005b: 73). I understand the intuition, especially if one takes into account that someone who benefits from an unjust structure may be less willing to change it. While we may thus be more worried about beneficiaries, it is not clear that benefiting itself makes them any guiltier than they would be if they just contributed to the structure without benefiting.

paying for adaptation – unless, that is, they are also those most able to help, in which case that factor – their ability to pay – is the reason they may be obliged to do so.

V Conclusion

I have argued that having benefited from past pollution does not, in and of itself, create duties to help those most vulnerable to adapt to climate change. We need additional arguments to make the case that benefits create such duties, and, once we have made these additional arguments, the question of whether anyone has benefited in some way or the other is no longer likely to be important. If one wants to maintain, for instance, that having benefited from a given action generates duties to provide aid to those who have fallen under a minimal threshold of well-being, it will have to be the case that having benefited coincides with being wealthy, absolutely speaking. But then it is the ability to pay and not having benefited that is doing the work. If one wants to claim, on the other hand, that having benefited from a given action generates duties of compensation, it is important to show that because of having benefited someone is failing in the duty to condemn a wrongful action or that, by benefiting, someone is directly implicated in producing a wrongful harm. Here again, it is not benefiting itself that gives rise to duties, but some form of responsibility involved in contributing to or perpetuating the harm.

For the case of past pollution this means that since there is arguably no claim to be made that benefiting from past pollution alone constitutes a separate offense or implicates the beneficiary in past pollution, the only reason to ask current beneficiaries to pay would be that they also happen to be the most able to pay – and this is, in fact, what frequently seems to be the case. In other words, I am not disputing the question of *who* should pay – I am only questioning a fairly commonly cited reason for *why*.

References

Birks, P. (2003). *Unjust Enrichment*. Oxford: Oxford University Press.

Butt, D. (2007). On Benefiting from Injustice. *Canadian Journal of Philosophy*, **37**(1), 129–52.

Caney, S. (2006). Environmental Degradation, Reparations, and the Moral Significance of History. *Journal of Social Philosophy*, **37**(3), 464–82.

Caney, S. (2010). Climate Change and the Duties of the Advantaged. *Critical Review of International Social and Political Philosophy*, **13**(1), 203–28.

Gosseries, A. (2004). Historical Emissions and Free-Riding. *Ethical Perspectives*, **11**(1), 36–60.

Kant, Immanuel (1797). The Metaphysics of Morals. In *Immanuel Kant. Practical Philosophy*, ed. Mary J. Gregor. Cambridge: Cambridge University Press (1996), pp. 353–603.

Meyer, L. H., and Roser, D. (2011). Climate Justice and Historical Emissions. In *Democracy, Equality, and Justice*, ed. M. Matravers and L. H. Meyer. London: Routledge, pp. 229–57.

Meyer, L. H. (2015). "Intergenerational Justice," *The Stanford Encyclopedia of Philosophy*, Edward N. Zalta (ed.), URL: http://plato.stanford.edu/archives/fall2015/entries/justice-intergenerational.

Miller, D. (2001). Distributing Responsibility. *The Journal of Social Philosophy*, 9(4), 453–71.

Miller, D. (2008). Global Justice and Climate Change: How Should Responsibilities Be Distributed? The Tanner Lecture on Human Values. Tsinghua University, Beijing, March 24–25. URL: http://tannerlectures.utah.edu/_documents/a-to-z/m/Miller_08.pdf.

Page, E. A. (2012). Give It Up for Climate Change: A Defence of the Beneficiary Pays Principle. *International Theory*, **4**(2), 300–30.

Pogge, T. (2005a). Recognized and Violated by International Law: The Human Rights of the Global Poor. *Leiden Journal of International Law*, **18**, 717–45.

Pogge, T. (2005b). Reply to the Critics: Severe Poverty as a Violation of Negative Duties. *Ethics & International Affairs*, **19**(1), 55–83.

Shue, H. (2010). Global Environment and International Inequality. In *Climate Ethics: Essential Readings*, ed. S. M. Gardiner, S. Caney, D. Jamieson and H. Shue. Oxford: Oxford University Press, pp. 101–11.

Singer, P. (2002). *One World: The Ethics of Globalisation*. New Haven, CT: Yale University Press.

Sinnott-Armstrong, W. (2010). It's Not My Fault: Global Warming and Individual Moral Obligations. In *Climate Ethics. Essential Readings*, ed. S. M. Gardiner. S. Caney, D. Jamieson and H. Shue. Oxford: Oxford University Press, pp. 332–46.

Welzer, H. (2008). *Klimakriege*. Frankfurt: S. Fischer Verlag.

6

Benefiting from Unjust Acts and Benefiting from Injustice

Historical Emissions and the Beneficiary Pays Principle

BRIAN BERKEY

I Introduction

If the worst potential effects of anthropogenic climate change are to be prevented, significant mitigation and adaptation efforts must be undertaken sooner rather than later. These efforts, if pursued, will be costly, and therefore one of the challenges that we face as philosophers interested in the normative implications of the threat that we face from climate change is attempting to determine the appropriate distribution of costs among those who are obligated to contribute. While most plausible views about which principles ought to determine the distribution of costs, and about which particular factors are relevant to the fair distribution of costs, tend to support the view that wealthy countries and individuals ought to shoulder the preponderance of the burdens, there are important differences among the competing views that have been advocated, and focusing on these differences is essential in order to adjudicate between plausible but incompatible theoretical positions.

Because plausible but incompatible views tend to have rather similar implications regarding the appropriate distribution of the costs of mitigation and adaptation efforts in the actual world, it can be helpful to consider what these views imply about the appropriate distribution of costs in circumstances that are in important ways different from those in the actual world. If a principle that seems independently plausible, and has implications regarding the appropriate distribution of costs in the actual world that seem at least roughly correct, nonetheless has implications regarding the appropriate distribution of costs in certain nonactual circumstances that seem intuitively unacceptable, this can provide a reason to reject the principle, and to seek an alternative that does not entail the counterintuitive implications in the nonactual cases. It is important, however, to avoid placing too much argumentative weight on intuitions about cases, whether actual or not, and too little on the independent plausibility of moral principles (Berkey, 2014: 160–63). So, while I will appeal to what seem to me to be counterintuitive implications of principles that have been defended by others in the course of arguing that we should reject those principles, I will also argue that there are important theoretical considerations that provide reasons to favor the type of view that I will defend over the alternatives that I reject.

An additional note about the method that I will employ in this chapter is that I will attempt to shed light on which principles ought to guide the distribution of the costs of mitigation and adaptation, and which factors are relevant to the appropriate distribution of those costs, on the assumption that it is possible to distribute the costs in accordance with the correct fundamental principles. My inquiry will proceed, then, at the level of ideal theory (Rawls, 1999: 8). Whether the correct fundamental principles can be appealed to in a way that will promote policy improvements in the real world depends on a wide range of empirical contingencies, and I will not discuss these here. In reasonably favorable circumstances, we have reason to hope that policy discussions and policy making in the actual world can be informed, at least to some extent, by philosophical reflection on the correct fundamental principles of justice. Of course, it may be that we do not find ourselves in such favorable conditions, in which case the relation between the basic principles that, on reflection, we endorse, and the manner in which we might engage in public debate over policy matters must become somewhat more complex. Still, I believe that even in rather unfavorable circumstances, our efforts to make the world less unjust ought to be informed by reflection regarding fundamental principles of justice (Simmons, 2010).[1]

My central aim is to provide some reasons to favor a view according to which neither historical emissions themselves, nor any relation that present individuals stand in to historical emissions or their effects, bear directly on the extent of the obligations that present people have to contribute to mitigation and adaptation efforts. I will use the term "historical emissions" to refer to emissions caused by or normatively attributable to people who are now dead. The view that I will defend does not imply, then, that currently living people who are responsible for substantial emissions that occurred earlier in their lifetimes do not have any special obligations, in virtue of their history of emitting behavior, to contribute to mitigation and adaptation efforts. Rather, I will argue that the fact that a person stands in some relation that others do not to the emissions of *other people*, and in particular of people who are now dead, whether it be in virtue of standing in some relation to the people responsible for historical emissions, to wealth produced via historical emissions, or any other similar relation that might be thought to be morally significant, does not, in itself, ground any special obligation to contribute to mitigation and adaptation efforts.

[1] For skepticism about this view, see Mills (2005); Farrelly (2007); Wiens (2012).

My argument, then, challenges the fairly widely held intuition that the history of behavior that has contributed to the threat of climate change that we currently face bears in a significant way on the obligations that current people have to contribute to mitigation and adaptation efforts (Baatz, 2013: 106; Caney, 2010: 214; Duus-Otterström, 2014: 448; Zellentin, 2014: 271). This intuition has led many to endorse versions of what have come to be known as the Polluter Pays Principle (PPP) and/or the Beneficiary Pays Principle (BPP), respectively.[2] The PPP holds that those who have themselves caused emissions, or are responsible for emissions, ought to bear the costs of mitigation and adaptation (Caney, 2005: 752; Caney, 2010: 204). Proponents of the PPP typically hold that polluters ought to shoulder costs in proportion to their level of emissions, and that only *unjust* levels of emission give rise to obligations under the principle. The more an agent has exceeded her fair share of emissions historically, then, the greater her fair share of the costs of mitigation and adaptation will be, according to the PPP.

The BPP holds that those who have benefited from emitting activity have a special obligation to bear costs of mitigation and adaptation, typically in proportion to the amount of benefit that they have received (Page, 2011: 420–21; Page, 2012: 306). Standard versions of the BPP hold that beneficiaries can be required, under that principle, to contribute up to the point that they are no better off than they would have been had they not benefited, but cannot be required to contribute more than that.[3] Like most proponents of the PPP, proponents of the BPP also typically hold that only benefits deriving from *unjust* historical emissions give rise to special obligations to contribute to mitigation and adaptation. Since it seems plausible that wealthy countries and wealthy individuals have benefited substantially more than others from unjust historical emissions, the BPP appears to provide a basis for assigning the majority of the costs of mitigation and adaptation to the wealthy, on grounds that are independent of the mere fact that they are wealthy.[4]

Despite the apparent appeal of both the PPP and the BPP, I will argue that neither provides acceptable grounds for assigning costs to present people on the basis of a relation in which they stand to historical emissions. I will discuss the PPP rather briefly in section II. The bulk of my discussion, which will occur in section "III, will focus on the BPP, since it seems to me the more plausible principle for assigning costs to present

[2] Some, such as Edward Page (2011), endorse both principles. Simon Caney, however, has argued that it is difficult to combine them in a plausible way (2006: 472–74).

[3] Proponents of the BPP generally allow that beneficiaries might be obligated to contribute more than the amount by which they have benefited to mitigation and adaptation efforts, but insist that any contributions that are required beyond the amount by which they have benefited must be required under a separate principle, such as the PPP or an Ability-to-Pay Principle (APP).

[4] This appearance is more difficult to defend than it might initially appear as a result of complexities raised by the Non-Identity Problem (Parfit, 1984: ch. 16). For helpful discussion, see Caney (2006: 474–46).

people on the basis of a connection to historical emissions. I will argue, however, that versions of the BPP that have been defended by others appear to share a common problematic feature. Specifically, they seem to limit the benefits that ground obligations under the principle to those that derive from *unjust acts*, and thereby implicitly deny that other ways in which individuals might benefit from injustice can ground similar duties to contribute to promoting justice. The versions of the BPP that I will criticize, for example, seem incompatible with the view that benefiting from *systemic institutional injustice* can ground special obligations to contribute to promoting justice, and with the view that benefiting from an *unjust state of affairs* that is not the result of unjust acts, in the sense of possessing more resources than one would possess in a just state of affairs, can ground such obligations.

The distinction between benefiting from unjust acts, on the one hand, and benefiting from systemic injustice or an unjust state of affairs, on the other hand, can be illuminated by looking at some key features of typical examples used by proponents to provide support for the BPP. These examples include the following:

- D attempts to divert water unjustly from B and C's land onto her own so that she can increase her crop yield, but inadvertently diverts the water onto B's land instead. As a result, B's crop yield doubles (B benefits from D's unjust act), C and D wind up with no crops, and D kills herself as a result (Butt, 2007: 132–33).
- At age fifty, you discover that you were admitted to Harvard only because of a bribe paid by your father, and learn the identity of the person who would have been admitted instead of you had the bribe not been paid. You are better off than you would have been had you not gone to Harvard, while he is worse off than he would have been had he been accepted (Goodin and Barry, 2014: 365).
- An enemy of my neighbor replaces a note that she left for her landscaper, whom she has prepaid, with a note instructing the landscaper to perform work on my yard that I wanted done but had not paid for yet. The landscaper performs this work, with the result that I have received free landscaping services while my neighbor has paid for services not performed (Butt, 2014: 338).

In all of these cases, the beneficiary of injustice innocently benefits from particular wrongful acts performed by others *within an institutional context*, rather than from the ordinary functioning of society's major institutions, or from an unjust state of affairs that results from the ordinary functioning of those institutions. Whether the relevant wrongful acts are performed within a just background institutional context, and whether the initial overall distribution of justice-relevant goods is just, are not specified. This suggests that those defending the BPP using these examples are operating on the assumption that the fact that one has innocently benefited from a wrongful or unjust act can generate reasons, and in some cases obligations, to redirect the relevant benefits to the victims(s) of the wrongful act, regardless of whether the background social and institutional conditions are just or unjust, and regardless of whether the beneficiary of the unjust act is herself a victim of systemic institutional injustice or an unjust state of affairs. Daniel Butt is explicit that his argument is focused on the question of "what a specified agent, who has benefited from *an instance of wrongdoing* which has caused a setback to the interests of another, should do" (2014: 337, italics added), and that he "put[s] to one side a range of problems about the extent to which one's actions should seek to bring society closer to one's preferred scheme of distributive justice" (2014: 337). The version of the BPP defended by Butt, then, like those defended by others, holds that one can be obligated to relinquish benefits acquired as a result of unjust acts performed within either a just or an unjust system, but takes no position on whether those who benefit from the operation of an unjust system can be obligated to relinquish the benefits that they enjoy as a result of its injustice.

It is unclear, however, what the theoretical basis might be for thinking that benefiting from unjust acts can generate potentially demanding obligations to sacrifice the relevant benefits in order to promote justice, while other possible ways of benefiting from injustice, such as benefiting from the operation of unjust institutions, do not generate similar obligations; and proponents of the BPP have not offered any argument in defense of this asymmetry.

The asymmetry might be defended by appealing to an account of justice, and in particular of distributive justice, that is at least primarily historical, in the sense that whether one is entitled to particular resources depends primarily on whether one acquired them in accordance with principles of justice for the acquisition and transfer of resources that apply to individual actions. This approach, however, is unlikely to appeal to most proponents of the BPP, since it would require endorsing an account of distributive justice that is at least much closer to right-libertarian views (e.g., Nozick, 1974) than they are typically inclined to accept. If, on the other hand, proponents of the BPP reject the asymmetry and hold that benefiting from injustice *of any kind* grounds obligations to give up the relevant benefits in order to promote justice, then on a wide variety of plausible views about distributive justice, historical emissions will not bear on the obligations of present people, since whether one is a beneficiary of injustice on the whole will not depend in any way on the relation that one's present holdings stand in to unjust historical emissions. Contrary to what its proponents have claimed, then, the most plausible versions of the BPP will not imply any significant role for historical emissions in the determination of the obligations of present people.

II The PPP and Historical Emissions

Proponents of the PPP believe that the *history* of greenhouse gas emitting activity, including emissions caused by people who are now dead, either directly or indirectly affects the extent of the burdens that ought to be assigned to current bearers of obligations to contribute to mitigation and adaptation efforts. One approach to defending this claim is to argue that we ought to treat countries, or nation-states, as the relevant bearers of obligations to contribute to addressing climate change (Miller, 2009: 121; Neumeyer, 2000: 186–89; Pickering and Barry, 2012: 670; Shue, 1999: 534, 545; Zellentin, 2014: 260, 265–68). This view, combined with the PPP, implies that the fact that some countries or nation-states have emitted more than their fair share historically provides a strong reason to allocate greater burdens to them than to those that have not emitted more than their fair share historically.[5] Since at least most wealthy industrialized countries have emitted more than their fair share historically, and at least most other countries have not, a version of the PPP that treats collectives such as countries or nation-states as the relevant agents for whom the principle generates obligations implies that, in the actual world, wealthy industrialized countries ought to shoulder most of the costs of mitigation and adaptation. This seems to be the right result, and the explanation that the PPP offers, namely, that those who have caused a problem have a special responsibility to pay the costs of addressing it, is familiar and, at least when applied at the level of individuals, quite compelling.

When we think about other cases, however, versions of the PPP that treat collectives as the relevant agents have implications that there are reasons to find troubling. Consider, for example, the following case, which I will call *Lost Wealth*:

Lost Wealth: Country A industrialized and emitted at very high levels beginning two hundred years ago, and as a result was quite wealthy. Beginning one hundred years ago, however, a combination of natural disasters, political mismanagement, and diminishing stocks of natural resources led to a sharp decline in economic productivity, and therefore a sharp decline in emissions. This trend has continued up to the present day. For the last hundred years the country's emissions have been a bit below its annual fair share,

[5] Neumayer refers to "countries" as the entities to be held responsible for historical emissions (2000: 186–87), as do Pickering and Barry (2012: 670). Zellentin claims that it is "states" that have obligations of rectificatory justice in virtue of being responsible for historical emissions (2014: 260, 265–68). Shue refers to both "countries" (1999: 534) and "states" (1999: 545). Miller rejects the view that rectificatory responsibility for historical emissions is most plausibly assigned to states, and claims that, instead, those who wish to assign to present members of a group responsibility for what previous members did should focus on "nations" (2009: 128). He suggests that while it may be unfair to insist that individuals shoulder burdens (e.g., tax burdens) because of what their states did in the past, it is not necessarily unfair to ask them to shoulder burdens on behalf of the national group with which they identify. This is because people with a shared national identity "think of themselves as belonging to the same cultural group as their forebears, and take pride in the historic achievements of their country" (2009: 128). He adds that "if you inherit the benefits of economic development, and claim the right to enjoy these benefits, by virtue of membership then you should also be held responsible for the associated costs" (2009: 128).

although its total historical emissions remain well above its fair share. Although the political situation in the country has improved a great deal in recent years, its economy remains weak for a variety of reasons, including a commitment by both the citizenry and the country's political officials to limit greenhouse gas emissions out of concern regarding the threat of climate change. As a result, current citizens of country A are, on average, significantly less well off economically than the citizens of most other countries that have emitted more than their fair share historically, and are also less well off than the citizens of many countries that have not emitted more than their fair share historically.

The PPP, applied to countries, implies at least that country A ought to bear a share of the costs of mitigation and adaptation that is a fair bit greater than its share would be if it were determined by present levels of wealth alone (i.e., if it were determined in accordance with an APP). In fact, the PPP, as it is often understood by proponents of applying it to collective entities, has even stronger implications regarding the obligations of country A. It is possible to combine the PPP with other principles, such as an APP, so that a country's appropriate share of the costs of mitigation and adaptation is determined by giving a certain amount of weight to its share of unjust historical emissions, and a certain amount to its present wealth. Many proponents of the PPP, however, view it as an alternative to the APP, rather than as a complementary principle (Neumayer, 2000; Zellentin, 2014), and others seem to believe that even if both historical emissions and present wealth should be taken into account, historical emissions should be weighted quite a bit more heavily (Page, 2011: 418–20). On these views, country A ought to bear a significantly greater share of the costs of mitigation and adaptation than other, much richer countries with lower historical emissions, and ought to bear a share of the costs that is not significantly lower than that of much richer countries with similar historical emissions.

Once we note that the costs assigned to a country will in fact be borne by its present citizens (Caney, 2006: 469), these implications seem quite troubling. The current citizens of country A had no control over the emitting behavior of those who lived in A 200 to 100 years ago (assume that no one in A is now more than 100 years old, or that those who are older are no more than 105 years old, so that they were very young children when the country last emitted more than its fair share in a given year), so it appears problematic to claim that they ought to shoulder the costs of what previous (and long dead) citizens did.[6] The PPP, applied

[6] Rudolf Schuessler presents a small-scale case that is roughly analogous, and that seems to me to effectively highlight the problematic feature of applying the PPP to collectives in the way that some of its advocates suggest. In his case, three children would, all else equal, be entitled to an equal share of a cake. The grandfather of one of the children, however, has wrongfully (but excusably, since he did not know that he was not entitled to any of the cake) eaten a slice. It seems unacceptable to hold that his grandchild should now receive a smaller piece of cake than the other children (Schuessler, 2011: 273).

to countries, can, then, unfairly burden those living in countries with high levels of historical emissions who are themselves not particularly well off. In addition, even when the PPP has implications that seem acceptable, it distributes burdens to people for the wrong reasons. If the fact that previous citizens of their country emitted heavily does not justify assigning substantial burdens to people who are relatively badly off, then that same fact cannot justify assigning substantial burdens to people who are well off either. In other words, if it is unfair to burden those who are not wealthy on the grounds that previous citizens of their country emitted more than their fair share, then it is unfair to burden those who are wealthy on the basis of *what previous citizens of their country did.* If the wealthy ought to bear substantial burdens, then, it must be for a different reason.[7]

The primary theoretical reason to reject the PPP, applied to collectives, then, is that it stands in tension with the highly plausible claim that individuals should not be obligated to bear greater burdens than others in virtue of nothing more than the behavior of other people over which they had no control. The intuition that the citizens of country A should not have to bear substantial burdens in virtue of what previous citizens of A did can be thought to support this general claim, but in fact the claim seems independently plausible enough that we might just as easily take the argumentative force of Lost Wealth to consist in simply highlighting the fact that the PPP, applied to collectives, can have implications that conflict with a theoretical claim that is quite independently plausible.

There is much more that can be said against versions of the PPP that take collectives to be the bearers of the relevant obligations, and I do not take myself to have provided an entirely decisive case against that view here.[8] I hope, however, to have provided sufficient grounds for doubting that it is acceptable to motivate those who are inclined to believe that historical emissions bear on present obligations to take themselves to have reason to consider alternative approaches to justifying that belief. The BPP represents one such approach, and it will be my focus in the following section.

III The BPP and Historical Emissions

Any version of the PPP that aims to allocate obligations to present people on the basis of historical emissions must take collectives to be the agents that bear the

[7] It can be a bit difficult to see this, since in the actual world there is such a strong correlation between a country's historical emissions and its present level of wealth. This correlation is, however, not perfect, even in the actual world (Caney, 2010: 212), and nonactual cases such as Lost Wealth help to make it clear that where this correlation exists, it is contingent, so that we must consider whether historical emissions have moral significance independent of present wealth, and if so, how much.

[8] Further discussion can be found in Caney (2006: 467–71).

relevant obligations.[9] The BPP, on the other hand, can assign obligations to either present collectives or present individuals on the basis of benefits received from historical emissions. Assigning obligations under the BPP to collectives, however, generates problems similar to those that I have argued make the PPP an unacceptable basis on which to assign obligations to present people.

Consider the version of the BPP endorsed by Edward Page, according to which "the burdens [generated by the BPP] should be distributed *amongst states* according to the amount of benefit that each state has derived from past and present activities that contribute to climate change" (2012: 302–03, italics added; see also Page, 2008: 563). Recall that in Lost Wealth, country A became quite wealthy as a result of its industrialization and high levels of emission between two hundred to one hundred years ago. Imagine that, in addition, the current citizens of A are no better off than (very likely different) citizens of A would have been had the country emitted only its fair share historically.[10] On at least most plausible accounts of what it is to benefit from unjust historical emissions, then, current citizens of A have not benefited from their country's unjust historical emissions. Imagine that they are also no better off than citizens of A would have been had *all countries* emitted only their fair share historically,[11] so that on at least most plausible views they have not benefited from *any* unjust historical emissions, whether those of their own country or those of others. Lastly, imagine that they are, on average, significantly less well off than the citizens of other historically high-emitting countries (whose current citizens enjoy at least a significant portion of the benefits that their countries have gained from historical emissions, whether those of their own country or those of others), and are less well off than the citizens of many countries that have neither emitted more than their fair share historically nor benefited from the historical emissions of others. Page's version of the BPP appears to imply that

[9] Recall that I use the term "historical emissions" to refer to emissions caused by or normatively attributable to people who are now dead. A version of the PPP that takes individuals to be agents of the relevant emissions, and therefore the bearers of obligations under the principle, would not, then, assign obligations to any present people on the basis of historical emissions (Caney, 2010: 210–11).

[10] Imagining how well off citizens of A would have been had the country emitted only its fair share historically entails imagining a (likely entirely) different set of current citizens of A. This is because a significant difference in A's level of historical emissions would affect who has children with whom, when people conceive, and so on, to such an extent that it seems at least fairly likely that none of the citizens of A who would exist in the circumstances described in Lost Wealth would also exist in the alternative circumstances in which A had emitted only its fair share historically (Parfit, 1984: 360–61).

[11] I use the notion of a country's "fair share" of emissions to refer to the amount X, whatever it might be, such that a country's emitting more than X (within the relevant period) constitutes an injustice. This is clearly the type of account of unjust historical emissions that defenders of the BPP typically have in mind when they claim that present people have benefited from unjust emissions and therefore have obligations under that principle to contribute to mitigation and adaptation. The basic idea is that past emitters unjustly (but perhaps nonculpably) appropriated an unfairly large share of the atmosphere's greenhouse gas absorption capacity for themselves or for their country, either because each person or country is entitled to a particular share of that capacity and no more, or because such appropriation violated a Lockean proviso requiring that appropriators leave enough and as good for others (Nozick, 1974: 178–82).

despite the fact that country A is currently not particularly well off, and despite the fact that none of its current citizens is responsible for any of A's unjust historical emissions, A, and therefore its current citizens, should be assigned a share of the costs of mitigation and adaptation that is at least a fair bit higher than the shares of similarly wealthy (and even wealthier) countries that have not benefited from historical emissions (either their own or those of other countries), and not significantly lower than the shares of countries that have benefited from historical emissions roughly the same amount overall over the past two hundred years, but whose current citizens (unlike the current citizens of A) enjoy a substantial portion of the total benefits received by their country.[12] This seems unfair to the current citizens of A for roughly the same reason that holding them responsible for costs under the PPP applied to countries is: they are held to be obligated to bear significant costs despite the fact that they themselves, as individuals, neither caused nor benefited from any unjust historical emissions. Since this seems unacceptable, we have reason to conclude that a plausible version of the BPP must take *individuals* to be the relevant beneficiaries for whom the principle generates obligations.[13]

Versions of the BPP according to which individuals are the relevant beneficiaries and bearers of obligations hold that insofar as an individual is a beneficiary of a relevant injustice, for example unjust historical emissions, she can be subject to obligations under the principle. Once again, the most commonly accepted versions hold that individuals cannot be obligated under the BPP to sacrifice an amount greater than the amount by which they have benefited. The particular implications of any version of the BPP will depend on what it counts as a relevant injustice – that is, what it counts as an X such that one's benefiting from X can give rise to

[12] Page's overall view about the appropriate distribution of the costs of mitigation and adaptation would exempt country A from bearing significant costs if it were below a certain threshold of current wealth (2011: 428). But so long as it is above that threshold, it appears that he is committed to allowing that it (and therefore its current citizens) can be required to bear costs that would lower it to the threshold, while other similarly well-off countries are obligated to bear only much lower burdens. This implication is troubling for the same reason that the implications of the PPP applied to countries is, namely, that it, at least indirectly, assigns substantial obligations to individuals on the basis of behavior of others over which they had no control, while assigning no obligations to currently similarly situated individuals.

[13] It might be suggested that a version of the BPP that takes countries to be the bearers of the relevant obligations could determine the extent of A's current obligations on the basis of the amount of benefit from unjust historical emissions that it *presently* enjoys. This would avoid unacceptably burdening A's current citizens, since A does not currently enjoy any benefits from historical emissions. But this view would not really treat countries themselves as beneficiaries and agents of obligations in a way that is morally consistent and analogous to how a plausible version of the principle would treat individuals. We would not allow a version that applies to individuals to, for example, exempt from obligations those who enjoyed benefits as a result of injustice in the past, but now no longer possess any of those benefits. If, for example, I nonculpably acquire a valuable item that turns out to be stolen, enjoy and benefit from its use for several years, at which point the item is damaged beyond repair, no plausible version of the BPP would assign me *no obligations* if the fact that the item was stolen is discovered only after it ceases to be a source of ongoing benefits to me. A version of the BPP that takes countries to be the relevant beneficiaries and agents of obligations, then, must take the fact that a country enjoyed benefits deriving from injustice *in the past* to bear on the extent of its present obligations under the principle. And, as I have argued, any view that does this will unfairly burden the present citizens of A.

obligations under the principle. The most common versions of the BPP include relatively narrow accounts of what can count as an X, according to which only unjust *acts* of various kinds can count. In principle, however, one might accept a version of the BPP according to which a wide range of types of injustice can be such that benefiting from any of them can generate obligations under the principle.

There is a powerful intuition that can be appealed to in support of the BPP, as applied to individuals. The intuition is that recognizing something as unjust commits one to recognizing reasons to remedy the injustice, if possible. If one has benefited from what one recognizes as an injustice, and can redirect the benefits that one has received to victims of the injustice, there does not seem to be any justification, all else being equal, for refusing to do so. This seems to be the case even if one had no way of avoiding receipt of the benefits in the first place.[14]

Despite the fact that a plausible way of explaining this intuition does not differentiate between types of injustice from which individuals might benefit, many proponents of the BPP have focused exclusively on benefits that individuals might acquire as a result of unjust *acts,* or, in some cases, patterns of acts (Baatz, 2013: 99; Butt, 2007: 143–44; 2014: 338–40; Gosseries, 2004: 50; Goodin and Barry, 2014; Haydar and Øverland, 2014).[15] On these views, the BPP gives rise to

[14] The intuition is nicely explained by Daniel Butt: "Moral agents can have obligations to compensate victims of injustice if they are benefitting and the victims are suffering from the automatic effects of the act of injustice in question… The individual's duty not to benefit from another's suffering when that suffering is a result of injustice stems from one's moral condemnation of the unjust act itself. In consequence, a duty to disgorge … the benefits one gains as a result of injustice follows from one's duty not to so benefit … taking our nature as moral agents seriously requires not only that we be willing not to commit acts of injustice ourselves, but that we hold a genuine aversion to injustice and its lasting effects. We make a conceptual error if we condemn a given action as unjust, but are not willing to reverse or mitigate its effects on the grounds that it has benefitted us. The refusal undermines the condemnation…Losses which others suffer as a result of the unjust actions of other persons cannot be dismissed as arbitrary or simply unfortunate: they create distortions within the scheme of fair distribution… If our moral condemnation of injustice … is to be taken seriously, it must be matched by action to remedy the effects of injustice" (2007: 143–44; see also 2014: 340). It is unclear, however, what Butt thinks the *conceptual error* one makes is supposed to be if she acknowledges that an action is unjust, but is not willing to give up benefits that she acquired as a result of that action. It seems quite plausible that refusing to give up the benefits amounts to a *moral* error, but it is difficult to see why we might think that it is also a conceptual error (Moellendorf, 2014: 170–71). Indeed, acknowledging that one has benefited from an injustice while asserting that one is nonetheless not obligated to give up the benefits that she has received in order to aid the victims of the injustice not only seems conceptually possible, but is in fact a widely endorsed view at least with respect to benefiting from systemic injustice. Most moral and political philosophers believe, for example, that those who benefit from unjustly low tax rates are not obligated to give up the additional amount that they should have been taxed in order to benefit those (e.g., the worst-off members of their society) who are the victims of the unjustly low tax rates and related failure of the state to provide income supplements or other benefits to them (for critical discussion of this view, see Cohen (2000: ch. 10)).

[15] Robert Huseby's (2015) criticism of the BPP also focuses exclusively on unjust acts. Two exceptions to this narrow focus are Holly Lawford-Smith's (2014) and Christian Barry and David Wiens's (2014) discussions. Despite endorsing a principle that clearly implies that the beneficiaries of various types of injustice, including systemic injustice, have obligations under the principle, however, the main example that Lawford-Smith uses in her defense of the principle is one in which a person benefits from normative failures constituted by particular acts or omissions (2014: 400–01). Barry and Wiens discuss both cases in which individuals benefit from unjust acts and cases in which they benefit from systemic injustice within companies, but they do not discuss the obligations of the beneficiaries of unjust political institutions directly. Nevertheless, the conditions under which they claim that individuals owe benefiting-related duties to victims of injustice appear to include

obligations that apply to those who have benefited from unjust historical emitting acts or patterns of such acts. The obligations that are generated by any particular version of a principle of this kind will depend on what counts as benefiting from unjust historical emitting acts. A range of plausible views about what counts as benefiting from unjust historical emitting acts, however, are such that combining them with a version of the BPP that takes individuals to be the relevant beneficiaries and bearers of obligations would have implications that there are reasons to find troubling.

Consider, for example, the view that what it is to benefit from unjust historical emitting acts is to be better off as a result of such acts than one would have been in their absence, holding everything else equal.[16] A version of the BPP that employs this account of what it is to benefit from unjust historical emitting acts would have either at least somewhat limited applicability to current individuals or implications that are much too sweeping. This is because of the nonidentity effect (Parfit, 1984: 360–61). At least some unjust historical emitting acts, and certainly patterns of such acts, affected which people came into existence, so that many people whom we might initially think are better off than they would have been in the absence of certain unjust historical emitting acts, in fact would not have existed at all had those acts not been performed. It would appear, then, that an individual cannot be said to have benefited from any unjust historical emitting acts that were necessary conditions of her coming into existence (Caney, 2006: 474–76). And since it is plausible that many individuals would not have existed if the unjust historical emitting acts from which we might have initially thought they benefited had not been performed, it seems that a version of the BPP according to which what it is to benefit from unjust historical emitting acts is to be better off than one would have been in the absence of such acts would not in fact generate obligations that are nearly as extensive as proponents may have thought it would.

Alternatively, it might be suggested that if certain unjust historical emitting acts were necessary conditions of one's coming into existence, then as long as one has a life that is worth living, one has benefited from those acts. But on this view individuals whose lives are barely worth living, and who possess none of the material goods or other advantages produced in unjustly emitting ways, would count as beneficiaries of unjust historical emitting acts, and so there would be reasons generated by the BPP for them to contribute to mitigation and adaptation efforts. These reasons might always be outweighed once other relevant considerations are taken into account, so that such people would not actually be obligated to contribute. But it is nonetheless implausible to hold that such

benefiting from, for example, unjust tax policies that leave them richer than they would have been under a just policy and others poorer than they would have been (2014: 14).

[16] This seems to be the view with which Axel Gosseries operates (2004: 49–53).

people are beneficiaries of unjust historical emitting acts in a way that provides any grounds at all (even if outweighed) for holding them responsible for bearing costs of mitigation and adaptation. These are certainly not people whom advocates of the BPP typically consider among the beneficiaries of unjust historical emitting acts.

A further problem with employing an account of what it is to benefit from unjust historical emitting acts according to which one has benefited if one is better off than she would have been in the absence of such acts can be seen by considering a variant of a case offered by Axel Gosseries (2004: 43–45) in his defense of a version of the BPP. In Gosseries's original case, the world consists of three countries, which I will call E, F, and G, respectively.[17] E emits at unjustly high levels, and F, while not itself emitting any greenhouse gases, benefits from the unjust emissions of E via extensive trade. G, meanwhile, is harmed by the unjust emissions of E because of the climate change caused by those emissions, and does not trade with either E or F (so it does not benefit from E's emissions in any way). Gosseries then imagines that E's population is completely wiped out by a natural disaster, so that it can no longer provide compensation to G for the effects of its unjust emissions. He claims that the citizens of F (strictly speaking, those individuals who have benefited from the unjust emissions of E) can be held responsible for providing compensation to the people of G, and that it is the fact that they have benefited from the unjust emissions of E that explains why they are obligated to provide this compensation.

Despite the fact that it might seem intuitively plausible that the citizens of F ought to provide aid to G in the circumstances described by Gosseries, it is not clear that the BPP provides the best explanation of this.[18] The fact that G does not trade with either E or F may, for example, leads us to assume that G is significantly poorer than F, so that the harm that it endures as a result of E's emissions exacerbates preexisting distributive injustice. In the absence of a presumption of this kind, or at least a presumption that the overall distribution of justice-relevant goods prior to G's sustaining harm and F's acquiring benefits as a result of E's unjust emissions did not unjustly favor G, the case seems under-described in ways

[17] Gosseries uses the names of actual countries, specifically the United States, the European Union, and Bangladesh. Because the use of actual country names may distort intuitive responses to the case, I have substituted generic names.

[18] As Robert Huseby (2015: 215) points out, the names of the countries used in Gosseries's original version of the example, along with the fact that G (Bangladesh) does not trade with either E (the United States) or F (the European Union), will lead readers to assume that citizens of F (the European Union) are at least mostly well off, while citizens of G (Bangladesh) are at least mostly poor. Since, as Huseby notes, "almost any plausible theory of distributive justice would demand assistance" from F (the European Union) to G (Bangladesh), it is difficult to see whether the fact that citizens of F (the European Union) have benefited from the unjust emissions of E (the United States) has any independent effect on the intuition that F (the European Union) should provide aid to G (Bangladesh) (2015: 215).

that undermine the possibility of generating reliable intuitions about whether F owes aid to G in virtue of benefiting from E's unjust emissions.

There are, in addition, reasons to doubt that the fact that citizens of F have benefited from the unjust emissions of E has any independent reason-providing force in Gosseries's case. Consider a variant of the case in which citizens of F are wealthy, F does not trade with E, and F is mildly negatively affected by climate change caused by E's unjust emissions. In this case, citizens of G, though very poor, benefit somewhat from the unjust emissions of E via trade, and G is not negatively affected by the climate change caused by those emissions. It seems implausible that the fact that G has benefited from the unjust emissions of E, while F has been harmed by those emissions, gives us reason to think that G should compensate F for the harms that it has suffered as a result of the unjust acts of citizens of E, once all of E's citizens have been wiped out by the natural disaster. The reason that this is the case seems clear. All things considered, citizens of G are victims of injustice, while citizens of F are plausibly beneficiaries of injustice. The fact that citizens of G have benefited from *particular unjust acts*, then, seems beside the point, even if citizens of F have not benefited from any unjust acts. Citizens of F are beneficiaries of systemic institutional injustice, since the global economic order unjustly favors their interests over those of citizens of G; and/or they are beneficiaries of an unjust state of affairs, in which they have more justice-relevant goods than they would have in a just state of affairs, while citizens of G have fewer. If one accepts that the global economic order is unjust in a way that benefits citizens of F, or that the state of affairs in which citizens of F are wealthy while citizens of G are very poor is unjust, then there do not seem to be any grounds upon which a proponent of the BPP could hold that the fact that citizens of G have benefited from the unjust acts of citizens of E generates compensatory duties owed to the citizens of the F, while the fact that citizens of F benefit from other forms of injustice does not generate duties owed by citizens of F to citizens of G. Of course one might deny that the global economic order is unjust, and that the state of affairs in which citizens of G are very poor while citizens of F are wealthy is unjust, and therefore claim that there is no injustice in the case that I have described apart from the unjust emitting acts of citizens of E. But I suspect that few proponents of the BPP would endorse this view, and it seems to me clearly unacceptable.

Proponents of the BPP might claim that the fact that citizens of G benefit from the unjust emitting acts of citizens of E provides some reason for them to compensate citizens of F for the harms that they suffer as a result of those acts, but that this reason is outweighed by the reasons provided by the fact that citizens of F benefit from other forms of injustice, so that on the whole citizens of F ought to provide aid to citizens of G in virtue of being overall beneficiaries of injustice while citizens of G are overall victims. This view, however, either does not actually give any independent weight to the fact that citizens of G have benefited from unjust acts, or else has implausible implications. To see this, suppose that the correct account of global economic justice is a sufficientarian account, and that the sufficiency level is set at twenty dollars per person per day. Initially, citizens of G earn fifteen dollars per person per day on average, so that E and F are jointly obligated to provide the equivalent of five dollars per person per day in aid. Citizens of G then benefit from the unjust emitting acts of E, so that they now earn sixteen dollars per person per day on average. The population of E is then wiped out by a natural disaster. It is true that had citizens of G not benefited from the unjust emitting acts of E, the citizens of F would have had to provide the equivalent of five dollars per person per day in aid, while they now, it seems, have to provide only the equivalent of four dollars per person per day. But this would be true regardless of how it came about that citizens of G are now better off than they had been. The fact that it was benefits resulting from unjust acts, rather than economic growth unrelated to any unjust acts, or any other cause, appears to make no difference to the extent of the obligations of the citizens of F. The fact that one party has benefited from unjust historical acts, then, appears to have no reason-providing force that is independent of the general requirements of justice.[19]

In order to hold that the fact that citizens of G have benefited from unjust historical acts has independent reason-providing force, proponents of the BPP would have to hold that because they have so benefited, citizens of F are obligated to provide less in aid to G than they would have been obligated to provide had citizens of G become better off in a way that was unrelated to any unjust historical acts. They might claim, for example, that in the case that I have described F is only obligated to provide three dollars per person per day in aid, rather than four dollars, so that citizens of G then have nineteen dollars per person per day on average. It is, however, implausible that a person's overall entitlements within a theory of justice that is not primarily historical could be reduced by the fact that she has become better off than she otherwise would have been (in an unjust system or unjust state of affairs) as a result of unjust historical acts. Rather, if there are independent facts about what constitutes a just outcome, or about what a just system and the results of its operation would be, then it is these facts that determine each person's overall entitlements. The

[19] The success of this argument does not depend on assuming that the correct account of global economic justice is sufficientarian. A similar argument could be made on the assumption that the correct account is, say, prioritarian or egalitarian. The relevant point is simply that for the fact that citizens of a country have benefited from unjust acts to have weight that is independent of the extent to which they are overall beneficiaries or victims of injustice, their having benefited from unjust acts would have to make a difference to their overall entitlements.

relation between one's current holdings and the actions of others over which one had no control (e.g., unjust historical acts), then, will not affect one's overall entitlements in any way.[20]

IV Conclusion

This line of reasoning suggests that we should take seriously a different type of principle that can be plausibly called a Beneficiary Pays Principle. Rather than holding that the beneficiaries of unjust acts have obligations under the principle, such a principle will hold that those who benefit from injustice in the sense that they possess more justice-relevant goods than they would within a just system or in a just state of affairs are obligated to transfer the justice-relevant goods that they would not possess in a just system or state of affairs to those who have fewer such goods than they would in a just system or state of affairs. Such a view would give no independent weight to the fact that one has benefited from unjust acts in the sense in which proponents of the BPP have typically understood what it is to benefit from such acts. It would also be, in important ways, more radical than the BPP as it has typically been understood. It would imply, for example, that beneficiaries of unjustly low tax rates have direct obligations to those who are the victims of unjust tax policy. And it would imply that if an unjustly poor person nonculpably receives money stolen from an unjustly wealthy person, the fact that the money was unjustly stolen has no independent reason-providing force that could generate an obligation applying to the poor person to return the money to the rich person.[21] Though many will find these implications counterintuitive, they seem to me correct, and I believe that a version of a Beneficiary Pays Principle that embraces them can be defended. Providing a thorough defense, however, is a task that must be left for another occasion.[22]

[20] If my argument here is correct, then it undermines not only the views of those who endorse versions of what they themselves refer to as a Beneficiary Pays Principle that focuses exclusively on unjust acts, but also the views of those such as Simon Caney, who rejects what he understands as the Beneficiary Pays Principle for several reasons, including issues relating to the nonidentity effect (2006: 471–76), but claims that in distributing the burdens of mitigation and adaptation, those whose wealth came about in unjust ways should, all else equal, be allocated a greater share than those whose wealth did not come about in unjust ways (2010: 217–18). On this view, two people with an equivalent amount of wealth should be allocated (perhaps highly) unequal shares of the costs of mitigation and adaptation if one inherited wealth created by a high-emitting company that employed his grandfather, while the other inherited wealth created by a company that did not engage in any unjust behavior. For reasons that I have already described, this seems to me to be an implausible implication, and it is one that is difficult to reconcile with an account of distributive justice that is not primarily historical in the way that libertarian views are historical.

[21] There may be other reasons that generate an obligation for the poor person to return the money to the rich person. But because the poor person is, by hypothesis, on the whole a victim rather than a beneficiary of injustice, despite being better off than she would have been as a result of a particular unjust act, she cannot, on the version of a BPP that I have suggested, be obligated *under that principle* to return the money.

[22] I am grateful to audience members at the George Washington University Department of Strategic Management and Public Policy, as well to Vince Buccola, Peter Conti-Brown, Nico Cornell, Gwen Gordon, Rob Hughes, Sarah Light, Eric Orts, Amy Sepinwall, and David Zaring, for helpful discussion. I thank two anonymous referees, as well as Lukas H. Meyer and Pranay Sanklecha, for helpful comments.

References

Baatz, C. (2013). Responsibility for the Past? Some Thoughts on Compensating Those Vulnerable to Climate Change in Developing Countries. *Ethics, Policy, & Environment*, **16**, 94–110.

Barry, C. and Wiens, D. (2016). Benefiting from Wrongdoing and Sustaining Wrongful Harm. *Journal of Moral Philosophy*, **13**, 530–52.

Berkey, B. (2014). Climate Change, Moral Intuitions, and Moral Demandingness. *Philosophy and Public Issues*, **4**, 157–89.

Butt, D. (2007). On Benefitting from Injustice. *Canadian Journal of Philosophy*, **37**, 129–52.

Butt, D. (2014). 'A Doctrine Quite New and Altogether Untenable': Defending the Beneficiary Pays Principle. *Journal of Applied Philosophy*, **31**, 336–48.

Caney, S. (2005). Cosmopolitan Justice, Responsibility, and Global Climate Change. *Leiden Journal of International Law*, **18**, 747–75.

Caney, S. (2006). Environmental Degradation, Reparations, and the Moral Significance of History. *Journal of Social Philosophy*, **37**, 464–82.

Caney, S. (2010). Climate Change and the Duties of the Advantaged. *Critical Review of International Social and Political Philosophy*, **13**, 203–28.

Cohen, G. A. (2000). *If You're an Egalitarian, How Come You're So Rich?* Cambridge, MA: Harvard University Press.

Duus-Otterström, G. (2014). The Problem of Past Emissions and Intergenerational Debts. *Critical Review of International Social and Political Philosophy*, **17**, 448–69.

Farrelly, C. (2007). Justice in Ideal Theory: A Refutation. *Political Studies*, **55**, 844–64.

Goodin, R., and Barry, C. (2014). Benefiting from the Wrongdoing of Others. *Journal of Applied Philosophy*, **31**, 363–76.

Gosseries, A. (2004). Historical Emissions and Free-Riding. *Ethical Perspectives*, **11**, 36–60.

Haydar, B., and Øverland, G. (2014). The Normative Implications of Benefiting from Injustice. *Journal of Applied Philosophy*, **31**, 349–62.

Huseby, R. (2015). Should the Beneficiaries Pay. *Politics, Philosophy, and Economics*, 14, 209–25.

Lawford-Smith, H. (2014). Benefiting from Failures to Address Climate Change. *Journal of Applied Philosophy*, **31**, 392–404.

Miller, D. (2009). Global Justice and Climate Change: How Should Responsibilities Be Distributed? *Tanner Lectures on Human Values*, **28**, 117–56.

Mills, C. (2005). Ideal Theory as Ideology. *Hypatia*, **20**, 165–84.

Moellendorf, D. (2014). *The Moral Challenge of Dangerous Climate Change: Values, Poverty, and Policy*. New York: Cambridge University Press.

Neumayer, E. (2000). In Defense of Historical Accountability for Greenhouse Gas Emissions. *Ecological Economics*, **33**, 185–92.

Nozick, R. (1974). *Anarchy, State, and Utopia*. New York: Basic Books.

Page, E. (2008). Distributing the Burdens of Climate Change. *Environmental Politics*, **17**, 556–75.

Page, E. (2011). Climate Justice and the Fair Distribution of Atmospheric Burdens: A Conjunctive Account. *Monist*, **94**, 412–32.

Page, E. (2012). Give It Up for Climate Change: A Defense of the Beneficiary Pays Principle. *International Theory*, **4**, 300–30.

Parfit, D. (1984). *Reasons and Persons*. New York: Oxford University Press.

Pickering, J. and Barry, C. (2012). On the Concept of Climate Debt: Its Moral and Political Value. *Critical Review of International Social and Political Philosophy*, **15**, 667–85.

Rawls, J. (1999). *A Theory of Justice*, rev. ed. Cambridge, MA: Harvard University Press.

Schuessler, R. (2011). Climate Justice: A Question of Historic Responsibility? *Journal of Global Ethics*, **7**, 261–78.

Shue, H. (1999). Global Environment and International Inequality. *International Affairs*, **75**, 531–45.

Simmons, A. J. (2010). Ideal and Nonideal Theory. *Philosophy and Public Affairs*, **38**, 5–36.

Wiens, D. (2012). Prescribing Institutions without Ideal Theory. *Journal of Political Philosophy*, **20**, 45–70.

Zellentin, A. (2014). Compensation for Historical Emissions and Excusable Ignorance. *Journal of Applied Philosophy*, **32**, 258–74.

7

A Luck-Based Moral Defense of Grandfathering

RUDOLF SCHUESSLER

The idea behind grandfathering is that the regulation of human practices should – at least to some extent – reflect the pre-regulatory state of those practices. In climate policy, this idea is usually implemented by basing emissions quotas on the historically established greenhouse gas emissions of countries. Governments regulate emissions trading with this approach, although most moral philosophers contest its legitimacy. In fact, a few years ago, Simon Caney observed that "no moral and political philosopher (to my knowledge) defends grandfathering" (Caney, 2009: 128). This position has changed in the meantime with Luc Bovens' and Carl Knight's elaborate moral arguments for grandfathering (Bovens, 2011 and Knight 2013 and 2014). Knight develops consequentialist and justice-based arguments, while Bovens ties grandfathering to Lockean appropriation and thus to issues of justice. I will also focus on justice here but reject the Lockean approach to demonstrate that the moral aim of alleviating bad luck quite naturally leads to grandfathering.

This chapter begins with an introduction to the concept of grandfathering, distinguishing variants of it and relating them to Bovens' and Knight's approaches (section I). Luc Bovens' account of Lockean grandfathering is discussed in section II and refuted (section III), however, not without seizing on a remark made by Bovens regarding the possibility of *temporary* grandfathering. My defense of temporary grandfathering builds on a luck-related but not luck-egalitarian approach to justice (section IV), and climate justice in particular (section V). The fundamental claim is that undeserved adverse shocks agents are subjected to ought to be buffered by a community that possesses the necessary means to do so. Above all, agents should be given time to adapt to an

I would like to thank the participants of the Graz conference and two anonymous reviewers for helpful comments. This research has been supported by the German Federal Ministry of Research, FONA grants 01UN1204E and 01UT1411C.

adverse situation that might otherwise cause severe suffering and disrupt their lives.[1] Temporary grandfathering is one way of meeting this demand by the residents of industrialized countries. There might be reasonable disagreement on the amount of acceptable time for adaptation. However, considerations concerning the social risks of rapid economic transformation (section V.1) and a reasonable speed of imposed habit change (section V.2) demonstrate that a short time frame is by no means self-evident. Moral considerations inform us about the configuration of an adequate adaptation process, suggesting gradual change and temporary grandfathering. Yet they do not specify the time frame of an adequate adaptation process provided that we allow for a reasonable amount of moral pluralism (section VI).

I Grandfathering

The term "grandfathering" derives from the legal notion of a "grandfather clause." The first and literal grandfather clause in U.S. law granted voting rights to all citizens whose grandfathers already had a vote (see Riser, 2006). Hence, African Americans were denied voting rights. In general, the term "grandfathering" now refers to regulations that reflect differences in power, endowments, or market shares or which already existed in a pre-regulatory state of affairs.[2] Grandfathering occurs in airspace regulations, railroad agreements, and distribution of CO_2 emission rights in Germany.

Problems related to emissions grandfathering have primarily been discussed by economists and lawyers in the past (see, e.g., Bohringer, 2005; Brandt, 2009; Weishaar, 2007). From a moral perspective, grandfathering is overwhelmingly rejected, though it has failed to attract the professional interest of moral philosophers, including those who specialize in climate ethics. This is most likely the result of a widespread equation of grandfathering with "squatter's rights" to high greenhouse gas emissions (see Neumayer, 2000: 188). The idea of "squatter's rights" suggests that the first occupier of a given resource acquires a property right to that particular resource. For many observers, this idea smacks of colonialism and the expropriation of indigenous peoples by European settlers. By analogy, historically high emitters have been accused of robbing the atmosphere's absorptive capacity from humanity as a whole. Such bleak analogies significantly blackened the image of emissions grandfathering, but can hardly be considered a substitute for a thorough moral analysis. Three recent articles have initiated a reconsideration of this status quo.

[1] Throughout this chapter, adaptation to new imposed conditions is defined in generalized terms. The distinction between adaptation and mitigation that has become customary in climate policy is not made here. Hence, mitigation measures are also adaptive in the present sense.

[2] Knight defines grandfathering with respect to GHG emissions as "the view that prior emissions increase entitlements to future emissions." This characterization is compatible with the one presented here, given that prior emissions occurred under no regulation of GHG output (Knight, 2013: 410).

Bovens and Knight have developed detailed moral defenses of grandfathering, or at least of some of its variants (Bovens, 2011; Knight 2013 and 2014).

Defenders of grandfathering respond to criticism of a "squatter's rights approach" by distinguishing among different forms of the doctrine. "Squatter's rights" and Lockean approaches in general base grandfathering on considerations of appropriation. Apart from such *property rights grandfathering*, the term "grandfathering" also refers to the calibration of emissions entitlements to a pre-existent level of output (as in the European Union). The entitlements are usually proportional to the emissions in a reference year. Such *reference point grandfathering* differs from its property rights cousin by telling us next to nothing about the reasons why grandfathering is accepted. For defenders of proportional emissions cuts who reject property rights grandfathering, this harbors the possibility of arguing in favor of their view on alternative moral grounds – an option that Knight embraces and that I will develop in a different direction in this chapter.

First, however, it should be noted that the question of appropriation is not the only moral bone of contention with respect to grandfathering. There is a huge difference whether we postulate temporary or lasting forms of grandfathering. In the former case, grandfathering diminishes with time and thus does not inform a final just distribution of emission entitlements. In the latter case, the distribution of emission entitlements – even in the long run – reflects historically acquired emission rights or the emissions in a reference year. *Temporary grandfathering* and *permanent grandfathering* clearly represent different claims, and we may expect that their moral justifications differ accordingly.

Bovens' and Knight's groundbreaking papers on the moral justification of emissions grandfathering can be assessed against this conceptual background.[3] Whereas Bovens defends a property rights approach, Knight adopts an instrumental and consequentialist perspective in his earlier article and includes a justice-based argument in his second article. However, both authors regard legitimate grandfathering to be temporary.[4] Hence, they do not use grandfathering to grant industrialized countries the permanent right to emit more than developing countries, but an entitlement to reduce emissions gradually. I agree that this claim best captures the moral significance of grandfathering and therefore also argue in support of temporary grandfathering, albeit on other grounds than Bovens or Knight.

[3] Knight uses a different terminology of grandfathering, principally distinguishing among strong, weak, and moderate forms of grandfathering. Strong forms assume permanent high levels of grandfathered emissions. Weak forms postulate grandfathering as a tie-breaker, that is, as holding when other moral considerations are on a par. Moderate grandfathering is *pro tanto* valid, but can be overruled by other moral considerations. The present double binary distinction is better attuned to the argument of this chapter (Knight, 2013: 411).

[4] Knight favors temporary grandfathering only contingently because in his instrumental approach, the balance of consequences determines when grandfathering ends, and hence will continue perpetually if it lastingly produces better consequences, as, however, is not to be expected. Bovens' view on the temporality of grandfathering is analyzed in sections II and III Approach".

Knight discusses several attempted justifications of grandfathering, settling in the end for one of its moderate forms. He rejects a pragmatic–egoistical approach (one could also say, naked power politics) and Bovens' Lockean reasoning, which favors instrumental arguments (Knight, 2013). In particular, Knight distinguishes between marginal cost and marginal benefit views. Grandfathering might be legitimate if high emitters faced particularly high abatement costs (ibid.: 416). However, Knight finds no good reason – and correctly so in my eyes – why this ought to be the case. He finds good reasons for the claim that an extra unit of emissions creates greater benefits in high- emission than in low-emission countries (ibid.: 418). It is important to note in this respect that industrialized countries could face economic breakdown if they were not given a sufficient package of emission entitlements to keep their economies running. Relative to such a scenario, the benefits of temporarily granting industrialized countries sufficient emissions for economic stability might be significant. This documents the importance of consequentialist considerations for the assessment of temporary grandfathering, even though the calculation of its benefits may be controversial. In the respective controversies, it is at least defensible to assume that prevention of economic breakdown in industrialized countries warrants a stepwise reduction of emissions. However, are we by endorsing this view sacrificing concerns of justice for the sake of consequences? Knight addresses the justice of grandfathering in his 2014 paper.[5] I will show here on a different basis that justice, and in particular justice in response to good or bad luck, does not stand in the way of temporary grandfathering. My argument starts with a critical review of Bovens' approach, which will serve as a counterpoint to the present approach.

II Bovens' Lockean Approach

Luc Bovens builds his moral philosophical defense of grandfathering on Lockean grounds (see Locke, 2003),[6] arguing that industrialized countries have – in a restricted but relevant sense – legitimately appropriated a larger share of the atmosphere's absorptive capacity during the process of industrialization. The classical tenets of Locke's theory of appropriation are used to justify this claim. The industrious and industrial front-runners of modernization have made good use of the atmosphere as a resource by creating progress and welfare not only for them, but for all. Locke's "no waste" condition may not always have been met, but to the extent that it was, the notion of inherited use rights of long-standing emitters is

[5] (Knight, 2014) appeared too late to be dealt with in detail in the present analysis, which was presented in 2011 and revised in 2013. Reference to Knight's justice-based claims will only be made where necessary to distinguish his approach from mine.

[6] Locke's and other classical theories of property have been discussed by several authors (Miller, 2009; Moellendorf, 2011; Posner and Weisbach, 2010; Risse, 2012; Starkey, 2011).

vindicated. Locke's "enough and as good" condition, which has only been satisfied up to a certain date in the past, appears more problematic. At present levels of greenhouse gas emissions, industrialized countries do not leave "enough and as good" of the atmosphere's absorptive capacity to others. Therefore, it is clearly inappropriate to extend the process of Lockean appropriation to the present. Nevertheless, a point in time must exist when the "enough and as good" condition was last met, that is, when the emission levels of the fast industrializers could still be universalized. It is this point in time (which I will call t^*) to which Bovens' Lockean considerations mainly refer. He argues that the relative emission shares at t^* should have some bearing on the distribution of emission rights in a coming international agreement on climate policy.

Bovens is aware of the objections that can be raised against this claim. He discusses and attempts to defuse three main objections to a Lockean approach. The first objection contends that Locke is concerned with private goods, whereas emissions control deals involve a public good. This objection ignores that Lockean arguments also apply to common pool resources, such as fisheries. I fully agree with Bovens that the private/public distinction is insufficient to rescind Lockean appropriation (and will say no more concerning this issue here).[7]

The second objection focuses on violations of the "enough and as good" condition and is, in my view, central to Bovens' claims. If it is determined that legitimate Lockean appropriation occurred up to t^*, why not allocate emission rights according to the emission quotas applicable to the emitters at t^*? New and rising emitters would then have to buy emission rights from established holders. Similar arrangements exist and are considered fair in other cases of common pool resources, such as fisheries. Fishing quotas are often set according to the annual catch ratios of fishing fleets when the unsustainability of further increases in yield is discovered. Bovens mainly alludes to such practices to undercut a quick transition to an egalitarian emission rights distribution. He points out that we do and should apply non-egalitarian standards in the case of fisheries and of fruit farmers, which in all moral respects are analogous to the climate case.[8]

A third objection against Lockean appropriation up to t^* arises from the structure of harm (see Bovens, 2011: 134). Violations of the "enough and as good" condition lead to harm in kind, for instance, depriving others of fish which they could have caught if overfishing had not occurred. The harmful effects of greenhouse gas emissions are often quite different: the sea will engross islands, such as Tuvalu; thousands of people will die because of droughts; and so on. Is this

[7] Actually, Locke's own reasoning is directly concerned with a common good case, because he assumes that God initially gave the earth to all human beings as joint property.

[8] "Do we say that everyone in the vicinity – fruit farmers or not, fishers or not – should now have equal access to the fruit-yielding capacity of the land or fish-yielding capacity of the lake and hence that larger operations should drastically downscale? I do not think so" (Bovens, 2011: 133).

a reason to reject Lockean appropriation? It seems clear that the harms in question should be prevented. However, preventing inacceptable harm to human societies and global ecosystems does not tell us how to distribute emission rights, since any distribution under an appropriately tight emissions cap will be compatible with the prevention of inacceptable harm of the kind described.

Hence, all three objections to Bovens' Lockean considerations seem to come to naught. He concludes:

> A radical egalitarian reform of emission rights without any concern for historically established claims [i.e., grandfathering; RS] is no less problematic than egalitarian land reforms without any concern for historically established claims.
>
> *(Bovens, 2011: 135)*

Against this background, it seems surprising that Bovens then turns around and declares that it "would be bordering on moral madness" to ask India to accept a 1:100 emissions per capita ratio relative to the United States – the ratio he probably assumes for t^*. It would be madness, indeed, if we were presently concerned with political agreement, but the question is a moral one: Why budge now, after having carefully established the moral case for a solution on the basis of relative emission shares at t^*? Bovens points out that any equitable emissions regime must rely on "multiple considerations." These considerations derive from an internal and external critique of the Lockean approach. The external critique asserts that humanitarian and utilitarian concerns should temper the use of property rights. It is important to note that this critique is not meant to confute the Lockean approach. It provides mitigating reasons, but in principle acknowledges Bovens' Lockean claims.

This also becomes apparent in Bovens' treatment of Nozick's example of a last well that provides water in a desert (ibid.: 137 and Nozick, 1974: 180). Nozick, who otherwise insists on the free use of property rights, admits that under certain conditions it would be immoral for a proprietor of water to refuse water to the needy. Nozick's specific conditions need not concern us here. It is more important that Bovens uses them to establish a *continued* "enough and as good" condition. If conditions of severe scarcity arise sometime after the legitimate appropriation of a resource so that suddenly far from "enough and as good" is available to others, their claim to the resource will resurface. The matter is further complicated because Bovens also claims that "some respect be paid to differential investments made during the time when there were no violations of the enough-and-as-good condition" (Bovens, 2011: 134).[9] On the other hand, asking developing countries to buy

[9] Moreover, Bovens considers Locke's "no waste" condition to be relevant: owners who have taken good care of a resource have a stronger claim to it, so that "if the no-waste condition is strongly respected, then the violation of the [continued] enough-and-as-good condition by itself is not enough to revoke my property rights" (Bovens, 2011: 140).

emission certificates is not a solution because they lack the required means to do so. In the end, a careful balancing of such reasons and counter-reasons documents that even Lockeans should not consider an emitter's historical emission record as the sole normative basis for the allocation of emission rights (ibid.: 141).

Bovens' conclusions reveal that he is not a diehard defender of "squatter's rights." He defends Lockean considerations only in terms of moral reasons, among others. Moreover, he argues that Lockean considerations should count most at the beginning of an emissions reduction process, while other reasons may phase in and even gain the upper hand later on. The result is *temporary* grandfathering, a temporary and waning recognition of higher emission quotas for historically high emitters. What Bovens most strongly inveighs against is, therefore, the immediate application of equal per capita emissions egalitarianism. However, his arguments do not preclude a vision of emissions egalitarianism as the final state of a global emissions regime.

III A Critique of Bovens' Approach

Bovens largely takes the sting out of property rights grandfathering by arguing that, all else considered, it will give rise to a temporary regime. On the other hand, his Lockean approach assumes that industrialized countries have legitimately appropriated more extensive emission rights than others. These rights might be counterbalanced by other considerations, but by virtue of Lockean principles, they remain rights. My critique focuses on this claim, arguing that industrialized countries never acquired property rights to particularly high emissions, above all not in a sense that would justify referencing to the emission totals of 1990 (the common reference point for proportional cuts). Although curtailing "squatter's rights" over time helps reduce the disparagement of permanent grandfathering, it neglects the fact that no significant "squatter's rights" exist in the first place. This can be shown with an analysis of the "enough and as good" condition of Lockean grandfathering.

The first step of Bovens' Lockean considerations is hardly disputable. If the "enough and as good" condition is violated, there has to be some (idealized) point in time t^* when the emission levels of industrializing countries no longer leave "enough and as good" for others. Since "enough and as good" refers to the atmosphere's absorptive capacity and climate science can only uncover the risks of different GHG levels in the atmosphere, t^* will in part depend on collective risk assessment. However, these considerations should not follow *our* current assessment of future climate risks, which willy-nilly takes the status quo of accumulated emissions into account. Lockean reasoning stipulates a state-of-nature scenario. In comparison with us, agents in a state of nature would probably prefer a considerably lower risk limit. From a state of nature perspective, agents can argue for zero tolerance toward climate risks, which means no temperature increase beyond the natural level. Some people in a state of nature might be willing to reap the benefits of industrialization. These individuals would probably accept a 1°C temperature increase, but I doubt that they would ex ante opt for 2°C. In any case, Lockean considerations in a state of nature have to posit a risk limit that is reasonably acceptable to all parties of a social contract.[10] Since people can reasonably disagree about the pros and cons of industrialization, the position of "no notable increase of greenhouse gas concentrations over preindustrial times" carries the day. The defenders of "no notable increase over preindustrial times" form a blocking coalition in the negotiations about a hypothetical social contract, and the negotiations will therefore fail or the contract will largely embody this position.

The upshot of these considerations is that *t** might have occurred very early in the process of industrialization. Bovens is of course right when he asserts that *t** must exist, because the climate can endure some positive amount of industrial greenhouse gas emissions. Nevertheless, it is highly doubtful whether a ratio of 100:1 between the United States and India would have existed at *t**. It would be fully compatible with the outlined natural-state reasoning if the only notable emitter at *t** were England (not even Great Britain), with the industries of France, Germany, and the United States either not yet existing or in their infancy stages. Parceling out emission rights according to emission shares at *t** would then create a monopolist (England) from whom all others would have to buy at their own cost. I take this to be a ludicrous assumption. In any case, my considerations concerning *t** are not intended to set a date. On the contrary, the wide gap between the earliest assumable date and Bovens' presumably late date indicates that we have no reliable estimate for *t** and therefore do not have a good reason to accept a specific ratio between the GHG emissions of major countries at *t**.

Yet even if we could identify *t** with some confidence, no justification for a projection of emission *quota* would ensue. In fact, it is not clear why successful Lockean appropriation in the climate case should be an appropriation of quota. A Lockean vindication of historical emissions merely implies that the amount of absorptive capacity consumed up to *t** (in tons of CO_2 equivalents instead of

[10] Appropriation on the basis of leaving "enough and as good" can, of course, proceed without a social contract or social approval. However, it is relevant that the interpretation of what is "enough and as good" is neither arbitrary nor fully individualistic for Locke. The interpretation of the proviso has to remain within the bounds of reasonableness to ground a moral title against interference by others, and in particular to ward off ex post nullification after a social contract has been concluded. (Such retrospective assessment of putative historical appropriations is in fact what matters from a Lockean perspective with respect to historical emissions.) Hence, reasonableness concerning emissions in a state of nature cannot be attained without taking into account what a community might ratify. On the role of reasonableness and epistemology for Locke's political philosophy, see, for example, Casson and Wilson (Casson, 2011: 92; Wilson, 2007).

emission shares) has been legitimately consumed. Therefore, nobody has to compensate others. This does not, however, imply that further consumption at an established rate is also legitimate on Lockean grounds. The absorptive capacity of the atmosphere for GHG is practically a non-renewable resource. The earth's ecosystem can to some extent recycle GHG, but this capacity does not suffice to offset the accumulation of GHG in the process of industrialization. Given the time scale of GHG decay, most GHG input over the last centuries is here to stay for the foreseeable future. Hence, the consumption of absorptive capacity is more or less analogous to the eating of a cake. Having eaten a fair share of the cake while others have just started nibbling at it does not entitle a fast eater to a larger piece of the cake, or to continued consumption at an established proportional speed relative to others.

The (near-) non-renewability of the atmosphere's absorption capacity for greenhouse gases over relevant time spans also implies a discrepancy with Bovens' land use or fisheries cases. Pastures or lakes generate new consumption goods in every period. As long as no overgrazing or overfishing occurs, there is a more or less plentiful flow of such new consumption goods. As already indicated, the atmosphere responds quite differently to the input of greenhouse gases, at least at the high level of input that became customary during the Industrial Revolution. Lockean considerations would thus have to focus on the appropriation of a practically non-renewable resource. Hence, Bovens draws a wrong analogy to the economy of renewable resources. Under these auspices, it is no longer plausible that industrialized countries should receive a privileged share of consumption after the limited nature of the resource has been discovered.

The analogy of the emissions case with pastures and fisheries also breaks down on a second front. Our willingness to accept the status quo as a yardstick to regulate fisheries resources, for instance, may hinge on other existing opportunities for wealth and welfare. In the European Union, for instance, some member states might readily acknowledge the dependency of others on fishing if their reliance on car production is reciprocally honored. As Bovens himself points out, such options do not exist in the emissions case.[11] If countries are deprived of the opportunity to achieve economic growth, which under the prevailing technologies is coupled to greenhouse gas emissions, they will hardly find other ways to boost the wealth and welfare of their citizens. Bovens uses this consideration to mitigate Lockean grandfathering, but in my eyes, his consideration strikes at its roots.[12] None of

[11] "Countries who do not have emission licences miss out radically in all aspects of life. ... The lack of industrialization within their borders keeps them in dire poverty" (Bovens, 2011: 141).

[12] Bovens' treatment of the Lockean proviso allows him to grant emission rights to the poor in cases of need by way of exception. However, I argue that no rightful Lockean appropriation of resources that others have a basic need for can even occur in the first place.

the disadvantaged countries can with good reason be asked to conduct its "pursuit of happiness" without emitting greenhouse gases. In this regard, a difference to the fisheries case exists, as new investors can reasonably be asked to turn to other markets.

The conclusion of such considerations is that Lockean rights (popularly "squatter's rights") to higher greenhouse gas emissions have no sound basis. Hence, it is fair to assume that industrialized countries did not acquire customary rights to emit more than others on a per capita basis. In particular, there is no reason to grant industrial countries higher emission rights permanently simply because they have a history of high emissions. Temporary grandfathering does not forestall this conclusion because it does not salvage Lockean grandfathering, which is always to some extent permanent.[13] Concessions to high emitters ought to be truly temporary at best, and we should therefore look for arguments that explain their truly temporary nature.

IV The Importance of Luck

Luck, chance, and fortune have a huge impact on how well our lives go. It stands to reason that ethics should take this fact into account. Lively debates on moral luck or compensation for bad luck show that contemporary ethics has risen to the challenge. Above all, luck egalitarianism has become a major theory of justice. As we will see, luck egalitarianism is not the only doctrine that helps us get a grip on the moral effects of undeserved bad luck. However, luck egalitarianism bears directly on matters of climate justice, and it is presently the most prominent approach to luck-related issues of justice. I will therefore begin this section with a few words on luck egalitarianism and its contribution to climate ethics.

Luck egalitarianism is concerned with the equalization of undeserved gains or losses in welfare, utility, human flourishing – or the opportunities thereof.[14] Its basic claim is that nobody should be in a better position than others without deserving that advantage. Hence, nobody should be better off because of an undeserved stroke of luck. Persons who are worse off because of "brute bad luck," that is, misfortunes for which they cannot be held responsible, should – according to luck egalitarianism – be compensated for their bad luck. If possible, their actual

[13] The example of Nozick does not compel us to think otherwise. The well owner's property rights are, in fact, permanent. Only with respect to the use of water, that is, the good that the well produces, will restrictions arise in times of drought or desertification (events producing what formerly was called "extreme necessity"). These restrictions limit the owner's free use of her property, but do not make others her co-owners. As soon as extreme necessity ceases, the owner regains the full use of her rights. Hence, the infringement of the owner's use rights is conditional and in principle temporary, whereas her acquired property rights are permanent.

[14] See Holtug and Knight for an overview (Holtug, 2007; Knight, 2009). I will not rigidly differentiate among various forms of luck egalitarianism in the following, because in one way or other, my contentions hold for all of them.

endowments or opportunities ought to be increased until all undeserved goods are equally distributed. Only merit or "option luck," that is, luck that results from morally imputable purposive action, is a legitimate reason for deviating from the aim of equal endowments or equal opportunities for all.

In climate ethics, this set of claims is usually understood as entailing duties of redistribution from high to low emitters of greenhouse gases (see Gosseries, 2007 and Neumayer, 2000). In fact, it is brute bad luck to be born poor in a poor country, which offers few opportunities for education or upward economic mobility to its poor citizens.[15] This state of affairs is causally linked to low carbon emissions. The wealth of nations has been fuelled by carbon emissions, and with prevailing technologies, any catch-up by underdeveloped countries will lead to rising carbon emissions.[16] A capping of global warming at 1–2°C above the preindustrial level calls for a tight emissions regime. Off-setting the brute bad luck of the poor in underdeveloped countries thus requires a redistribution of emissions permits from presently high to low emitters (there are, of course, also other reasons for a redistribution of emission shares).[17]

Obviously, these considerations undercut the rationale for grandfathering. If grandfathering is to be backed by luck-related moral arguments, they will apparently have to be based on different lines. Yet it does not take much to see that luck egalitarianism can also be adduced in favor of high emitters who have to reduce their GHG output substantially. It is often acknowledged that the leaders and citizens of industrialized countries were long inculpably ignorant of the climate effects of GHG emissions. Necessary awareness of these effects began around 1990, when news of global warming was a surprise to the leaders and citizens of industrialized countries. Awareness of anthropogenic global warming entailed a responsibility to reduce GHG output considerably, and given the potential or even expectable negative welfare effects of such reductions, learning about global warming at a time when advanced economies already heavily depended on fossil fuel is to be regarded as a stroke of brute bad luck.[18]

[15] Of course, being born poor in a rich country is also brute bad luck, not least because it implies an undeserved bad starting position for life. However, this predicament is an issue for the internal justice of rich societies, which have the means to alleviate it on their own. Moreover, the claims of the poor in rich countries do not nullify the claims of the poor in poor countries.

[16] It must be borne in mind, of course, that the riches of industrialization did not come about as mere windfall profits or simply as a result of fossil fuel burning. It took much thought and effort to create the wealth of nations. How these contrary considerations weigh in in the final judgment will not concern us here. Let us just keep in mind that luck egalitarianism provides some compelling arguments for a redistribution of emission shares from rich to poor countries.

[17] Note that luck egalitarianism is not synonymous with approaches that burden the beneficiaries of industrialization or those with an ability to pay. See Meyer and Roser (2010) for the beneficiary pays principle (Meyer, 2004; Meyer and Roser, 2010). Caney argues for a modified Ability-to-Pay Principle (Caney, 2010).

[18] I will revert to this line of reasoning in section V.

What then does luck egalitarianism demand if undeserved bad luck befalls rich people (or even ordinary citizens of industrialized countries)? If brute bad luck causes significant losses for persons who are well off, they might become poor and need support as a matter of justice for the undeservedly poor. This is, of course, not the expected case in climate policy. It is instead to be expected that the citizens of industrialized countries might lose some advantages while still remaining better off than the global average. Luck egalitarianism seems to call for a distinction in such cases. Apparently, nothing needs to be done against windfall losses of undeservedly advantaged persons as long as they retain above-average welfare or opportunities. In such cases, luck is merely undeservedly taking away what luck has undeservedly given. It makes a difference, however, if the advantaged deserve their wealth (because of hard toil, etc.). Now, a rectification of brute bad luck seems in order, although the well-off person may still be better off than the average. This is because any deviation from a just, merit-based distribution caused by brute bad luck stands in need of rectification according to luck egalitarianism. Hence, deservedly rich persons who encountered losses because of a stroke of brute bad luck deserve compensation out of the undeserved resources of possibly much poorer persons.

It is not immediately clear whether a defense of grandfathering follows from this claim. Much depends on the extent to which industrialized nations deserve their wealth and on the nationality of the resource providers. Many of them would presumably be from industrialized countries so that no net increase in the emission entitlements of industrialized countries ensues. This suggests that we look elsewhere for a justification of emissions grandfathering. A further reason for this suggestion results from the underlying claim itself, which appears utterly counterintuitive to me. Why should we assume that the windfall losses of deservedly rich billionaires (which leave them billionaires) require compensation out of the purses of ordinary people, even if these people acquired their modest wealth by good luck? Luck egalitarians may have other intuitions with respect to this question, but all those who consider a positive answer to be inappropriate have reasons not only to deny compensation to the billionaires but also to reject luck egalitarianism.

I will accept this conclusion and apply a different set of luck-regarding assumptions. Let us assume that everybody is entitled to basic resources and opportunities in life (in a sufficientarian manner, if you like).[19] Gains and losses above the level of basic endowments are subject to the vicissitudes of luck. Hence, if a person

[19] Sufficientarianism roughly holds that considerations of distributive justice are valid up to a threshold of wealth, welfare or opportunities (see, e.g., Frankfurt, 1987; Shields, 2012). They do not support claims of redistribution above that threshold. My present considerations link up with sufficientarianism if brute bad luck below the sufficiency threshold is treated differently from brute bad luck above it. However, I do not need to hold specific views concerning a sufficiency threshold here. In particular, I do not assume that the level of basic endowment is the sufficiency level.

undeservedly loses resources or opportunities above the basic level, the person's own resources ought to be used to cover the loss and its consequences. There are two cases in which a person's own resources may not cover the loss. First, the resources may simply be insufficient to prevent the person from falling into poverty (i.e., the person ends up with less than her basic endowment). In this case, the entitlements of the poor come into play. Second, the person's resources may suffice to ward off poverty, or the shock may not lead to impoverishment in the first place. In these cases, a problem might still exist. The crucial point is that losses can cause suffering even if they leave a person well off in the end. Losses often require adaptation to a new after-loss situation, and the process of adaptation can be very painful. The prevention of serious human suffering is one of the primary values in ethics. Luck-regarding ethical theories should therefore grant support to persons who undeservedly undergo processes of imposed painful adaptation. This does not compel us to help people who are themselves responsible for the adaptive pressures they face, at least as long as they remain above the level of poverty. I will here for simplicity's sake side with the view that such individuals are not entitled to help and concentrate on cases of undeserved, fortuitous adaptive pressures.[20]

Note that a title of support against adaptive shocks does not imply that people ought to be fully shielded against fortuitous adaptive pressures above the level of basic endowment. I assume that people are responsible for adapting to new situations to the extent that adaptation is a reasonable demand. Hence, if a person prefers not to adapt although adaptation is a reasonable demand, she or he has to bear the consequences. Any claim to support is only a safeguard against excessive burdening. In the present context, this means that others should help the agent to adapt. They need not conserve a prior status quo.

Of course, much more deserves to be said about the outlined assumptions, but I will presently not attempt a full-fledged critique of luck egalitarianism. My premises may not be acceptable to luck egalitarians, but they should appeal to those who aspire to a less demanding, and in my eyes more realistic, luck-regarding morality. On this premise, it is possible to formulate a *buffering principle*:

> People who suffer greatly under adaptive pressures through no fault of their own ought to receive the resources or the time they need to adapt in a morally acceptable way if they lack the means to do so on their own. Acceptability implies that, if possible, the suffering of those people should not exceed a threshold that is to be determined by moral considerations.
>
> *(The costs of intervention may enter these considerations.)*

[20] It seems plausible to me that even self-incurred suffering ought to be buffered above a certain level of suffering, and that this level is higher than in the case of luck-induced suffering. Yet I will not pursue this issue here.

The buffering principle is a principle of justice for the transition from one state of the world into another. Such processes of transition are manageable in just or unjust ways – a fact that moral philosophers, who otherwise concentrate on abstract patterns of distributions, should not disregard. Considerations of buffering are of course not the only moral reasons that may govern a transition. Therefore, the buffering principle has a *pro tanto* status. Still, helping people digest undeserved adverse shocks and giving them time to adapt are among the most plausible aims of luck-related ethics. Social policies that buffer the undeserved fall of people highlight the attractiveness of this aim. Such policies usually operate under the assumption that people have to adapt and that helping them adapt or stretching the time for adaptation is often the only remaining option.

Note that the buffering principle is compatible with elements of luck egalitarianism, but cannot be integrated into luck egalitarianism. We might claim, with whatever plausibility, that the costs of a social buffer against brute bad shocks should be borne by the holders of undeserved benefits. However, the buffering principle stands in conflict with luck egalitarianism because it emphatically refuses to equalize the effects of brute luck. On the contrary, it accepts that human lives are subject to the influence of fortune, which morality in general need not rectify but mitigate at best, unless people fall short of a basic set of resources and opportunities. This implies that bad brute luck is mitigated with respect to a person's own prior status rather than levelled with respect to intersubjective differences in resources or advantage. The former characterizes the buttressing approach, whereas the latter is typical of luck egalitarianism.[21] This difference also distinguishes the present approach from that of Knight (2014), who argues that grandfathering can be justified on the basis of egalitarian principles.[22]

Does it matter whether adaptation starts from a deserved or undeserved position? Imagine the case of Antoinette, who was born into a rich and noble family. She grew up in considerable wealth and acquired most of her noble peers' consumption habits. One black Friday her family loses all their wealth without any fault of Antoinette's. She forthwith has to earn her living as a chambermaid. Should society buffer Antoinette's fall and help her adapt to her new life as a

[21] I owe this observation to a referee. The buffering principle also differs from luck egalitarianism with respect to its approach to expensive tastes. If brute bad luck renders the fulfillment of expensive tastes impossible, for which a person bears no responsibility, luck egalitarians believe that the person deserves compensation (see Knight, 2009: 16n 46; Knight, 2014: 577). In contrast, the buffering principle only demands that we help the person in question to adapt her preferences, and only if the person cannot pay for herself. Hence, support for adaptation should not be proportional to the strength of an expensive taste but to a person's ability to adapt without inordinate suffering.

[22] Knight also argues that his analysis with respect to egalitarianism carries over to prioritarianism and sufficientarianism (Knight, 2014: 582). Since his claims concerning the latter theories of justice indirectly depend on his treatment of egalitarianism, they also differ from the present approach.

chambermaid?[23] I think that Antoinette's case is morally not much different from one in which the agent has herself earned her wealth. Hence, the buffering principle ought to be applied. What matters for the buffering principle is that Antoinette's habits were not immoral and that the acquisition of these habits occurred through no fault of her own and through no particular risk taking on her part. We therefore owe support to Antoinette, who, of course, nevertheless did not deserve to be born richer than others. It should be clear that buffering does not require Antoinette to be allowed to continue to live on her previous level of wealth. What reasonable buffering could be in her case is up to a moral community. Maybe we should temporarily provide her with the resources to meet her old friends once in a while. Maybe we should pay for psychotherapy if Antoinette turns out to be addicted to her expensive tastes or her old environment and falls into a deep depression under the pressure to adapt. I do not want to speculate about such measures; the point is that Antoinette ought, to some extent, to receive help to adapt.

Finally, there is one type of suffering that is of particular interest here. People should primarily be buffered against brute bad luck that destroys their lives. Shocks of fortune can tear a life to shreds, and such shocks not only cause severe suffering but often also deprive a person of the psychological means to adapt. Examples of such events are severe illnesses, sudden unemployment, the breaking apart of personal bonds, and so on. Let us call an event life disrupting if it disrupts a person's normal flow of life so that life no longer feels complete. Since the buffering principle's objective is to safeguard adaptation, it becomes doubly important to buffer people against brute shocks that deprive them of adaptive potential. In other words, we should take special care to ensure that people do not suffer life-disrupting shocks.

V Brute Bad Luck Meets Western Lifestyles

What do such considerations concerning buffering teach us about climate ethics? Extensive emission cuts in industrialized countries will probably force ordinary people with ordinary tastes to change their way of life. Even if the possibility of emissions trading is taken into account, the old consumption habits will become costly and their owners might often suffer under the pressure of adaptation. Some Porsche drivers will presumably have to change to a less expensive car because they can no longer afford the petrol costs. Many of us may not care much about such problems. However, rising energy prices will probably negatively affect poor

[23] I implicitly assume that society has the means to help Antoinette, that these means are not earmarked for more needy persons, and so on.

and average citizens in industrialized countries much more than the Porsche drivers. Some poor people may hardly be able to afford a car, and ordinary people will have to struggle to make ends meet if they keep their cars. Ordinary people will have to pay significantly higher rents for well-insulated, energy-saving houses. The resulting reduction in mass mobility or the curtailing of other lifestyle options will very likely hurt many people in industrialized countries. Of course, optimistic estimates of climate mitigation costs assume relatively low welfare losses for industrialized countries. However, short-term cost-curve studies of various energy scenarios, which are more reliable than long-term cost–benefit calculations, have a less optimistic outlook (see Enkvist et al., 2007).[24] Nobody should underestimate the task of achieving 80 percent or 95 percent emission reductions in industrialized countires by 2050. This is what climate experts are calling for, and if we take their demands seriously, we will have to live on a budget of approximately two to three tons of CO_2 equivalents per person and year. It only seems reasonable to expect that coping with this budget will prove painful for many.

It is important to recognize that these considerations do not express the perspective of an overall aggregative consequentialist approach in the present context. Of course, allusions to losses, pain, and life disruption are references to consequences. But as a justice-oriented approach, buffering implies that it is not – or not only – the balance of consequences that matters.[25] Imposing or failing to buffer life-disrupting shocks can be unjust, even if the overall balance of consequences in a society is positive. This is particularly the case if the shocks negatively affect already disadvantaged or vulnerable members of a community, or render others vulnerable. Considerations of buffering can hence inveigh against a duty of rapid adaptation, even if adaptation only costs a small overall loss of Gross National Product.[26]

There is, of course, also the question of responsibility for global warming. People who are responsible for the non-tenability of their privileged position arguably cannot demand buffering if their privileges cease to exist. Such a refusal of support may even be in order if the fall from grace is a sudden shock. Those who

[24] Some economic cost–benefit analyses of climate policies, such as the Stern report, predict rather low welfare losses in industrialized countries (Stern, 2007). If these analyses are correct, the adaptation costs of a greening economy are hardly worth discussing. Given the well-known shortcomings of economic modeling as a prognostic tool, it is, however, not clear how much trust we should put in such long-term calculations. For a criticism of the Stern report, see Tol (2009).

[25] There is a further consideration showing that buffering and overall consequentialist calculations need not lead to the same results. Note that costs and losses are only indirectly relevant for buffering. The citizens of industrialized countries may have to accept rising monetarized opportunity costs of a green lifestyle. The viability of an economic transformation without breakdown, deep crisis, and life disruption is the yardstick for economic adaptation, not monetarized cost or BIP losses as such.

[26] Those who insist on aggregative consequentialist moral guidance instead of a multi-criteria approach should at least integrate the costs of buffering (through welfare schemes) into their economic calculations. To the best of my knowledge, this has hitherto not been done in climate economics.

culpably ignore that an adverse shock might afflict them do so at their own risk. However, one of the most important claims in climate ethics is that the citizens of industrialized countries were *inculpably* ignorant about the greenhouse effect or the impact of greenhouse gas emissions on the earth's atmosphere during most of the emission histories of their countries.[27] If inculpable ignorance about global warming in the past is acknowledged, the time when Western citizens first ought to have become aware (whether they actually knew or not) of anthropogenic global warming attains a crucial significance. Like many moral philosophers, I accept that such a point in time exists – or rather that the gradual historical accumulation of information about global warming should be mapped on an idealized point in time t_{alpha} when the scope of this information became too great to be inculpably ignored. The year 1990 is often accepted as t_{alpha} because it roughly corresponds to the founding of the IPCC (1988), the first IPCC report (1990), and the agreement on a Framework Convention on Climate Change (1992). It does not matter for our theoretical purposes here whether you prefer an earlier or later date, but I will use the conventional reckoning and assume $t_{alpha} = 1990$. Under this premise, the citizens of industrialized countries experienced (or would have experienced if they complied with their epistemic duties) a brute information shock in 1990.

At first glance, it seems appropriate in such a situation to adopt the principle that justice ought to prevail as quickly as possible.[28] This means that we should implement a valid demand of justice as quickly as physically possible. The downside of such a claim is obvious. It neglects the costs of implementing justice and in extreme cases instantiates the principle "Let justice be done, even if the world perishes" (*fiat justitia, pereat mundus*). Moral philosophers of many schools reject this principle. Yet even if ultimate doom is not at stake, the buffering principle tells us that the citizens and economies of industrialized countries should be given an appropriate amount of time to adapt to greener lifestyles. This is not merely a pragmatic claim. It goes beyond recognizing veto power and giving industrialized countries more than the physically minimal time for emission cuts as a political compromise. Moreover, buffering claims do not depend on an overall consequentialist calculation. They show that the speed of economic and ecological adjustment to a sustainable state is subject to antagonistic claims of justice. On the one hand, we have the buffering principle, but on the other, the entitlements of the world's poorer citizens are pushing in an opposite direction. Such antagonistic claims need pondering in moral discourse and public deliberation. It is very likely that public deliberation will not resolve disagreement about the appropriate weighing, and different reasonably defensible positions can be entertained with

[27] On inculpable ignorance and historical emissions, see Caney and Schuessler (Caney, 2006; Schuessler, 2011).
[28] Müller argues for rapid end-state justice (Müller, 2009).

respect to the expected results. Simple economic balancing cannot solve this impasse, because different considerations of justice are involved. On a practical level, each step toward an agreement will have to be negotiated.

It is very probable that a policy of gradual adjustment in industrialized countries will assume the form of temporary grandfathering. True to the buffering principle, rich countries will have to bear the costs of ecological adaptation themselves but may request a suitable time schedule for adaptation. Temporary grandfathering means that pre-existing emission levels (of, say, 1990) are accepted as a reference point for the reduction of GHGs. It is difficult to see how life disrupting shocks on the way to a greener economy in industrialized countries could be forestalled without giving due weight to existing emission levels. Sections V.1 and V.2 will spell this out in more detail. Moreover, percentage cuts to GHG emissions are the most straightforward way to achieve a gradual mitigation of climate change. The weight of antagonistic considerations can then be accommodated by demanding higher or lower percentages for the cuts in a given period or by shortening the period for given cuts. Under these auspices, we can claim that temporary grandfathering is a fair approach to emission reduction, buttressed by the buffering principle of luck-regarding justice.

Of course, it remains unclear how to strike a balance between the moral reasons or justice claims involved. This is a warning not to expect too much guidance from ethics in climate negotiations. As indicated, a moral compromise needs to be negotiated, too. However, it seems helpful to keep some general considerations in mind. It is clear that the buffering principle will justify less buffering today than in 1990. After all, citizens and economies of industrialized countries ought to have adapted to a good extent by now. Those who refused to adapt in time did so irresponsibly (above all in democracies) and do not deserve the benefits of undiminished buffering. Buffering is only appropriate to the extent that agents who fulfill their moral duties still deserve it at a given point. It ought to rely on a scenario in which all requested steps of adaptation were taken, thus creating a path of proper adaptation against which deviations can be measured. The path of due adaptation defines what is up to the agent and what should be considered as excess burden. Many of those who are worried about global warming will think that the path of adaptation for industrialized countries ought to have ended by now. This is tantamount to claiming that full ecological adjustment is overdue. Such claims depend on the weighing of antagonistic moral reasons, and a plurality of reasonably defensible claims may therefore exist (we usually disagree about the relative weight of moral reasons). However, since many highlighted reasons in the literature on climate ethics speak in favor of a brief period of grandfathering, it may be worthwhile to expound countervailing reasons for an extended period of grandfathering as well.

V.1 Changing an Economy

Two contexts of buffering provide particularly strong reasons for a protracted period of grandfathering: the first concerns the functioning of social institutions and economic systems; the second entails the adaptation of human habits to a GHG-minimizing lifestyle. This section addresses the first context; the discussion of habits follows in the next section.

The buffering principle calls *pro tanto* for the prevention of life-disruptive shocks (if necessary by buffering the shocks), and it thus becomes a demand of justice to give economies – or, more precisely, the agents who operate under given economic institutions – time to adapt. Some justice-related claims and ecological necessities call for a shortening of the suggested time span, but on the other hand, economic limitations to rapid systemic change actually commend an extension. There is disagreement concerning the feasible speed of adaptation to a fully sustainable economy. We need not take sides in this dispute because it suffices for our purposes to postulate that major economic change (on which I will focus in the following) cannot occur in a very short time without life-disrupting shocks for most economic agents involved. In this regard, all those who suggest massive emission cuts ought to specify whether they would also accept a drastic rise in unemployment, a rapid and severe lowering of income, or a loss of social security in the wake of adapting to a greener economy. (This is a reminder that considerations that favor the buffering of the worst afflicted persons are irreducible to overall cost–benefit calculations.) The answers to this question are important for determining the timetable of an adjustment of industrialized economies to a final fair regime of GHG emissions. Opinions concerning the necessary time will presumably differ, but the hardships a claimant is willing to accept with respect to transition should be in the open. In any case, moral considerations do not seem to exclude societies from opting for a course of adjustment that implies only a low or moderate risk of such hardships. Universally compelling moral standards for prescribing a more painful course are lacking. It follows that no country can be unilaterally morally required to risk economic collapse (from a mainstream political and economic perspective) if it cannot even be sure that its sacrifices will engender a sufficient global mitigation of GHG because of the uncertainties of international cooperation.[29]

Finally, it should be noted that the role of economic transition for the buffering of brute bad luck differs significantly from the role of economic endowments in Lockean accounts of appropriation. Bovens argues that legitimately acquired property deserves protection so that established stocks of property should only be devaluated slowly (see Bovens, 2011: 134). Yet property is always at risk in

[29] The fact that high emitters may be jointly responsible for reducing GHG output does not tell us much about each emitter's obligation to act despite disagreement about climate policies.

capitalist economies.[30] Why protect stocks of property against ecological shocks but not against market shocks? This apt question shows how important it is to recognize that buffering considerations are not concerned with the protection of property. They protect people against brute bad luck, whether it originates in the natural environment or in the market. Since property or income is usually a prerequisite of an undisrupted life, luck-regarding ethics will temporarily buffer the loss of property or income. However, this is not the upshot of a theory of property and does not amount to a principled title to a gradual devaluation of property. The present approach is therefore not Lockean or libertarian.

V.2 Changing Habits

It seems natural to assume that only the "generation of 1990" in industrialized countries can request buffering with respect to the brute bad luck of having to change their lifestyle or ingrained energy intensive habits. Later generations ought to know right from the start that CO_2-intensive lifestyles are no longer sustainable. Note that it does not matter what these people actually believe about global warming. The epistemic duties of ordinary people imply a duty of information gathering, and after 1990, a proper evaluation of the available information concerning global warming must lead to the insight that a "business as usual" attitude is no longer acceptable. Under this premise, the risks of adopting a CO_2-intensive lifestyle after 1990 are transparent and therefore option luck, that is, luck for which an agent is fully liable.[31] It apparently follows that post-1990 generations cannot claim to be negatively affected by climate-related lifestyle changes. The buffering of informational brute bad luck seems appropriate for the generation of 1990 only.

However, this rejoinder proves inadequate on closer inspection. Large parts of post-1990 generations in industrialized countries still become accustomed to a CO_2-intensive lifestyle during the formative years of their youth. They may acquire (or at least learn about) ecological values at school and at home. However, when they reach the age at which they make decisions about their own lifestyles, a difficult situation arises. On the one hand, conscientious reflection tells them (or ought to tell them) to adopt an eco-sensitive lifestyle; on the other hand, some CO_2-intensive consumption preferences already became entrenched in their youth. This means that even post-1990 generations face brute bad luck with respect to no longer sustainable preferences, because at the time when a conscious decision is due, losses arise relative to entrenched preferences for which the agents are not

[30] I thank the participants at the Graz conference for this objection.

[31] For a detailed analysis of the legitimacy of expectations concerning unsustainable life styles, see Meyer and Sanklecha (Meyer and Sanklecha, 2011, 2014). For a discussion of Meyer and Sanklecha's complex approach, see Culp (2011).

responsible. The fact that – other than for the generation of 1990 – the information about necessary change is not new for later generations does not impugn this result, because the relevant point of reference is the point in time when members of these generations become responsible for their use of information, not when they first received it.

Moreover, scientific certainty concerning the human impact on climate has further increased since 1990. A focus on 1990 helps to systematize arguments and to analyze aspects of climate policy and climate ethics. However, it should not blind us to the fact that the knowledge in question spread gradually among scientists, political decision makers, and the public at large. As certainty concerning these facts grows, it becomes increasingly difficult to reject the demand for lifestyle change reasonably. There is a difference between the first time when a probable hypothesis impinges on responsible action planning (in order to ward off risks) and the time when it has to be accepted as a fact that no reasonable person can deny. This difference in time translates into a difference in cost and efforts that have to be borne in response to a claimed fact. In other words, the justifiable amount of effort to mitigate greenhouse gas emissions has grown since 1990 and therefore the adaptive cuts reasonable persons have had to accept became larger. Hence, it would be wrong to restrict the informational impact of brute luck to the reference point of 1990. If the first bad newswas brute bad luck, successive confirmation was more of it.

These considerations call for a gradual approach to lifestyle changes, especially when accounting for the risk of life disruption. Extraordinary steps of emission reduction entail the risk of demolishing the life-world of individuals. It is therefore crucial not to overestimate the speed with which ecological best practice can spread in a society and consequently become a demand on average citizens. Such worries, of course, dovetail with institutional and technological problems of economic transformation. Both sets of considerations for an extended period of grandfathering therefore reinforce each other.

VI Conclusion

This chapter has examined two approaches to the moral justification of emissions grandfathering: Bovens' Lockean approach and a luck-based "buffering" alternative. Bovens' approach has been scrutinized and found wanting. A Lockean approach would lead to historically established "squatter's rights" of high emitters, even if further moral considerations may mitigate and qualify such rights. Bovens' arguments fail for two reasons. First, Lockean emission quotas depend on a time t^* when still "enough and as good" emission capacity was left for all of earth's inhabitants. As shown, emission quotas at t^* cannot be determined with any confidence. Hence, the Lockean argument cannot support an assignment of emission rights at levels that are associated with grandfathering. Second, Bovens' analogies to the economy of renewable resources are misleading because the depletion of the atmosphere's capacity for greenhouse gas emissions is practically nonrenewable given the established scale of output. Consequently, the historical emission levels of industrialized countries do not engender customary rights to further high emissions. Against this background, it does not suffice to constrain the "squatter's rights" of high emitters because no such rights are justifiable in the first place.

However, the arguments against property-rights grandfathering do not carry over to temporary reference-point grandfathering. We should therefore take an unprejudiced look at this form of grandfathering. I have presented a luck-based moral argument for the temporary acceptance of proportional emission cuts (proportional to historically established emission levels at a reference point such as 1990). My argument relies on a buffering principle that gives people time to adapt to a new way of life in cases of severe undeserved losses. Hence, not only the poor in poor countries but also the citizens of industrialized countries, who have to develop ways of sustainable living, are entitled to help, in particular by setting a suitable time frame for proportional emission cuts. This timetable should not be determined on the basis of economic cost–benefit calculations alone, because buffering is a demand of justice. Hence, buffering the disadvantaged and vulnerable in industrial countries attains a peculiar role beyond a mere balance of economic costs and benefits. I have argued that not only the generation of 1990 is entitled to gradual adaptation, but to some extent so are subsequent generations. The first subsequent generations will still be accustomed to unsustainably high consumption levels before they begin to assume moral responsibility for their own lives. They also have to find a compromise between the risk of economic breakdown and a speedy transition to a greener mode of production.

The upshot is a conception of "contraction and convergence" that rests on two pillars.[32] The first pillar consists in a fair end state of emission rights distribution independent of historically grown emissions.[33] The second pillar is an irreversible process of continuous GHG output contraction for presently high emitters up to the end state. However, the timescale of this process cannot authoritatively be determined on moral grounds. Its duration is negotiable, not only for reasons of political necessity but also because morality[34] does not provide uncontentious reasons for setting a timetable.

[32] For the "contraction and convergence" approach, see Meyer (2000, 2007).

[33] See Schuessler (2013) for an argument for emissions egalitarianism as an apt end state of emissions right allocation.

[34] I refer here to standard ethical theories and theories of justice. There is also a morality of making compromises in negotiations, but I will not venture into this field here to find out whether it – unexpectedly – contains a timetable for grandfathering.

References

Arnold, D. (2011). *The Ethics of Global Climate Change*. Cambridge: Cambridge University Press.

Bohringer, C. (2005). On the Design of Optimal Grandfathering Schemes for Emission. *European Economic Review*, **49**, 2041–55.

Bovens, L. (2011). A Lockean Defense of Grandfathering Emission Rights. In *The Ethics of Global Climate Change*, ed. D. Arnold. Cambridge: Cambridge University Press, pp. 124–44.

Brandt, U. (2009). The Choice between Auctioning and Grandfathering in the EU. *Energy and Environment*, **20**, 1117–30.

Caney, S. (2006). Environmental Degradation, Reparations, and the Moral Significance of History. *Journal of Social Philosophy*, **37**, 464–82.

Caney, S. (2009). Justice and the Distribution of Greenhouse Gas Emissions. *Journal of Global Ethics*, **5**, 125–46.

Caney, S. (2010). Climate Change and the Duties of the Advantaged. *Critical Review of International Social and Political Philosophy*, **13**, 203–28.

Casson, D. (2011). *Liberating Judgment: Fanatics, Skeptics, and John Locke's Politics of Probability*. Princeton: Princeton University Press.

Culp, J. (2011). Comment on Lukas Meyer and Pranay Sanklecha. *Analyse & Kritik*, **33**, 473–76.

Enkvist, P., Nauclér, T., and Rosander, J. (2007).A Cost Curve for Greenhouse Gas Reduction. *McKinsey Quarterly*, **1**, 35–45.

Frankfurt, H. (1987). Equality as a Moral Ideal. *Ethics*, **98**, 21–43.

Gosseries, A. (2007). Cosmopolitan Luck Egalitarianism and Climate Change. *Canadian Journal of Philosophy Supplementary*, **31**, 279–309.

Holtug, N. (2007). *Egalitarianism*. Oxford: University Press.

Knight, C. (2009). *Luck Egalitarianism*. Edinburgh: Edinburgh University Press.

Knight, C. (2013). What Is Grandfathering? *Environmental Politics*, **22**, 410–27.

Knight, C. (2014). Moderate Emissions Grandfathering. *Environmental Values*, **23**, 571–92.

Locke, J. (2003). *Two Treatises of Government*. New Haven, CT: Yale University Press.

Meyer, A. (2000). *Contraction and Convergence*. Dartington: Green Books.

Meyer, A. (2007). The Case for Contraction and Convergence. In *Surviving Climate Change: The Struggle to Avert Global Catastrophe*, ed. M. Levene and D. Cromwell. London: Pluto Press, pp. 29–58.

Meyer, L. H. (2004). Compensating Wrongless Historical Emissions of Greenhouse Gases. *Ethical Perspectives*, **11**, 20–35.

Meyer, L. H., and Roser, D. (2010). Climate Justice and Historical Emissions. *Critical Review of International Social and Political Philosophy*, **13**, 229–53.

Meyer, L. H., and Sanklecha, P. (2011). Individual Expectations and Climate Justice. *Analyse & Kritik*, **33**, 449–71.

Meyer, L. H., and Sanklecha, P. (2014). How Legitimate Expectations Matter in Climate Justice. *Politics, Philosophy, and Economics*, **13**, 369–93.

Miller, D. (2008). Global Justice and Climate Change: How Should Responsibilities be Distributed? The Tanner Lecture on Human Values. Tsinghua University, Beijing, March 24–25. URL: http://tannerlectures.utah.edu/_documents/a-to-z/m/Miller_08.pdf.

Moellendorf, D. (2011). Common Atmospheric Ownership and Equal Emissions Entitlements. In *The Ethics of Global Climate Change*, ed. D. Arnold Cambridge: Cambridge University Press, pp. 104–23.

Müller, O. (2009). Mikro-Zertifikate. *Für Gerechtigkeit unter Luftverschmutzern. Archiv für Rechts- und Sozialphilosophie*, **95**, 167–98.

Neumayer, E. (2000). In Defense of Historic Accountability for Greenhouse Gas Emissions. *Ecological Economics*, **33**, 185–92.

Nozick, R. (1974). *Anarchy, State, and Utopia*. New York: Basic Books.

Posner, E., and Weisbach, D. (2010). *Climate Change Justice*. Princeton: Princeton University Press.

Riser, V. (2006). Disfranchisement, the U.S. Constitution, and the Federal Courts: Alabama's 1901 Constitutional Convention Debates the Grandfather Clause. *American Journal of Legal History*, **48**, 237–79.

Risse, M. (2012). *On Global Justice*. Princeton: Princeton University Press.

Schuessler, R. (2011). Climate Justice: A Question of Historic Responsibility? *Journal of Global Ethics*, **7**, 261–78.

Schuessler, R. (2013). *Equal Emissions per Capita: A Moral Defense*. Discussion paper, University of Bayreuth, Germany.

Shields, L. (2012). The Prospects for Sufficientarianism. *Utilitas*, **24**, 101–17.

Starkey, R. (2011). Assessing Common(s) Arguments for an Equal per Capita Allocation. *The Geographical Journal*, **177**, 112–26.

Stern, N. (2007). *The Economics of Climate Change: The Stern Review*. Cambridge: Cambridge University Press.

Tol, R. (2009). The Stern Review: A Deconstruction. *Energy Policy*, **37**, 1032–40.

Weishaar, S. (2007). The European Emissions Trading System and State Aid: An Assessment of the Grandfathering Allocation Method and the Performance Standard Rate System. *European Competition Law Review*, **28**, 371–81.

Wilson, C. (2007). The Moral Epistemology of Locke's 'Essay'. In *The Cambridge Companion to Locke's "Essay Concerning Human Understanding,"* ed. L. Newman. Cambridge: Cambridge University Press, pp. 381–405.

8

In Defense of Emissions Egalitarianism?

CHRISTIAN BAATZ AND KONRAD OTT

I Introduction

The idea that every person is entitled to the same amount of greenhouse gas (GHG) emissions has gained momentum. It receives broad support from academics, non-governmental organizations, and leading politicians. Given that, according to this view, emissions are to be distributed in an egalitarian fashion, in the following we will refer to Emissions Egalitarianism (EE). In his book *One World – The Ethics of Globalization*, Peter Singer (2002) has famously argued for EE and in past writings we have endorsed this position as well (Ott et al., 2004; Ott and Baatz, 2012).[1]

Criticism of EE usually seems to be motivated by self-interest of high polluters. Derek Bell (2008) and Simon Caney (2009), however, criticize EE for being "Atomist" and "Isolationist": EE falsely ignores other climate-induced costs, such as those resulting from adaptation, and EE ignores other considerations regarding global justice, such as poverty.[2] Bell and Caney provide forceful arguments against EE. This chapter aims at investigating whether and to what extent EE can be defended against this criticism.

We start with a brief reconstruction of the global commons argument for EE and discuss whether or not the atmosphere – or rather, the earth's sink capacity to absorb GHG – can be classified as a global commons (section II). Section III then briefly outlines how we will proceed in the reminder of this article.

II The Argument for Emissions Egalitarianism

Singer (2002: 28) starts his argument for Emissions Egalitarianism with the claim that the earth's atmosphere is a global common good, a claim that is endorsed by

[1] EE have also been adopted, amongst others, by the German Advisory Council on Global Change (WBGU, 2009) and Anil Agarwal and Sunita Narain (1991).

[2] Others have argued along similar lines, for example Baer et al. (2007), as well as Johan Eyckmans and Erik Schokkaert (2004). See also the very recent books by Moellendorf (2014) and Shue (2014).

many (see, e.g., Vanderheiden, 2004; WBGU, 2009). Christian Seidel (2012, 2013) reconstructs the global commons argument for EE in the following way:

(P1) The atmosphere is a global commons.
(P2) A global commons is owned by everyone equally.
(P3) If a global commons is owned by everyone equally, the right to use it should (*pro tanto*) be distributed equally amongst all.
(C) The right to use the atmosphere should (*pro tanto*) be distributed equally amongst all.

In our opinion, the reconstruction accurately captures the key steps of the global commons argument for EE.

Before discussing several challenges to this argument, some clarifications regarding (P1) are required. It is obvious neither what exactly the good is that we are concerned with nor how to classify it. Starting with the former, different terminologies are employed throughout the literature: in addition to "the atmosphere," the respective common good is referred to as "climate," "climate change," the "absence of climate change," or "climate stability" (see Seidel, 2012: 182).

The good that we care about and want to sustain is a stable climate which provides beneficial boundary conditions for flourishing human (and animal) lives. The term "stability" is a concept that entails a multitude of beneficial climatically induced states of affairs. A stable climate is threatened by overuse of the earth's sink capacity. On the other hand, it is always possible to add more GHG molecules to the atmosphere. Thus, one must be more precise: what is limited is the earth's capacity to absorb GHG without climate change or without "dangerous" climate change (given that anthropogenic climate change already takes place). Still, we do not know with certainty how much GHG can be added to the atmosphere before causing "dangerous" climatic changes. We must rely on probabilities instead. Seidel (2013) therefore proposes that the respective good can be defined as follows: "the capacity of the [earth] to absorb GHG emissions such that the risk of morally unacceptable climate change lies below *p*", or alternatively "a level of atmospheric GHG concentrations such that the risk of morally unacceptable climate change lies below *p*".[3] Note that the good defined in this way contains a normative element (in contrast to "the atmosphere/climate/sink capacity", etc.): what qualifies as "dangerous" or "morally unacceptable" climate change has to be determined by ethical reasoning; a value judgement is required (see Ott et al., 2004).

Is "the earth's capacity to absorb GHG emissions such that the risk of morally unacceptable climate change lies below *p*" a global commons, as claimed by

[3] *p* is a probability.

Singer? A global commons is non-excludable but rival in use: when it is provided to some, others cannot be prevented from consuming the good as well (non-excludability), and consumption of the good by some constrains the consumption of others (rivalry).[4] This seems to be correct, for only a limited amount of GHG can be absorbed without risking morally unacceptable climate change. That is to say, once a cap on total emission output is introduced, consumption is rival. But what about the non-excludability condition? In the absence of regulatory schemes, anyone can use the earth's sink capacity (open access). If a scheme is introduced that governs GHG emissions, however, it is possible to limit an agent's access to the sink. At least on the level of individuals it is not possible to exclude agents completely from emitting GHG (assuming they cannot be prevented from breathing and digesting). This seems to indicate that the respective good is not a *pure* global commons because others can be excluded to use the good to a certain extent under a regulatory scheme.

Moreover, the sink capacity that we have focused on so far is only one of two relevant aspects. If unrestrained access to the earth's sink capacity is successfully limited, a new good is realized: a protected or stable climate, that is, a climate system without morally unacceptable climatic changes (Seidel, 2012: 184). This good will be used as well all around the globe, insofar as people benefit from the absence of certain negative impacts of climate change, for example extreme events. Consumption of *this* good is non-rival and non-excludable. Others cannot be prevented from benefitting from a stable climate[5] and consumption of some does not inhibit the consumption of others. The fact that this good will benefit different people unequally constitutes neither rivalry nor non-excludability. Hence this good is a (pure) public good, for both conditions (non-excludability and non-rivalry) are met (Seidel, 2012: 184, 187). In contrast to "classic" public goods such as radio programs, the good contains a normative component, however (see preceding discussion).

Apparently, climate change and the prevention of climate change, respectively, have two relevant aspects with regard to global goods: the earth's sink capacity and a stable climate. Although both aspects provide different kinds of benefits, they are interrelated: the more the sink capacity is used, the less stable the climate system will become.[6] There is a causal driver (human-induced emissions) that impairs this stability over time. Overutilization of a global commons (sink capacity) impairs a global public good (stable climate). Thus, we face a "dialectical good". Depending

[4] The classic example of a global commons is a commonly owned sheep run.

[5] Sometimes others *can* – in principle – be excluded from benefitting from a stable climate by relocating them to an area without these benefits. Given the very high costs associated with such measures we think that in practice excluding others is hardly possible. Seidel (2012: 183) discusses this issue.

[6] As Seidel (2012: 184) puts it: "Commons and pure public goods are closely linked inasmuch as the prevention of overuse of a commons always creates a pure public good" (our translation).

on the aspect on which one focuses, the good has to be classified differently, either as a (impure) global commons or as a pure global public good. Note that both aspects occur in both definitions given previously: the phrase "such that the risk of morally unacceptable climate change lies below *p*" indicates the stable climate aspect, and "the earth's capacity to absorb GHG emissions" and "a level of atmospheric GHG concentrations" both indicate the sink aspect (for sink capacity in GHG concentrations are interrelated).

In sum, (P1) should be specified in the following way: the atmosphere (as shorthand for Seidel's definitions) is a (impure) global commons *and* a global public good.

III Procedural and Conceptual Clarifications

Now, what about (P2) saying that "a global commons is owned by everyone equally" given that the atmosphere is this dialectical good? Can it still be characterized as a good that belongs to everyone equally? We think that it can. Both global commons and global public goods are owned by everyone equally. Why, then, should this hybrid or dialectical good that contains elements of both goods not be owned by everyone equally? Even if one concludes that the sink capacity is not adequately classified as a commons, why should the good as defined here not be owned by everyone equally? One may justify unequal ownership by drawing on Locke's argument for the appropriation of natural resources that are owned by no one (*res nullius* instead of *res communis*). Locke's overall argument, however, does not support a *res nullius* position. Locke argued that, originally, the bountiful earth has been given to humankind in general. Acquisition of natural resources by means of labor and ownership can be justified if a proviso will be respected. Locke's (1993) famous proviso requires that "enough" in terms of both quantity and quality should be left for others. Locke's proviso is clearly violated regarding the appropriation of the earth's sink capacity (see Bovens, 2011; Schuessler, 2011).

However, one could further question (P2) by claiming that instead of the atmosphere the totality of all global commons is owned by everyone equally (see Caney, 2012: 268–71). Assuming for the sake of argument that this claim is correct, it would result in a very complex allocation procedure. The reason is that distributing use entitlements of the atmosphere must then take into account how (benefits from) all other global commons and public goods are distributed. We have grave doubts regarding the practicality of such a proposal. Anyway, these concerns are structurally similar to our discussion of (P3), and we will conclude that (P3) is false. That is to say, for the sake of the argument we will assume that (P2) is correct. Still, given that (P3) fails, the global common good argument cannot justify EE. After reaching this conclusion we will, however, argue that something in defense of EE can be said on "practical grounds". For these reasons we will not discuss (P2) in any detail but turn directly to the more controversial (P3).

Critique of (P3) can be mapped onto the distinction between Holism/Atomism, on the one hand, and Integrationism/Isolationism, on the other hand, that was recently introduced by Caney. In a nutshell, the Holism–Atomism distinction refers to whether the distribution of costs generated by different strategies to respond to climate change should each be dealt with separately (Atomism) or whether overall climate-induced costs should be distributed in a combined fashion (Holism). The Integrationism–Isolationism distinction refers to whether the distribution of climate-induced costs should ignore the current distribution of overall burdens and benefits (Isolationism) or whether it should be taken into account (Integrationism).[7]

(P3) is plausible only within an approach that is Atomist *and* Isolationist. In the following, we will show why this is so. We will first discuss arguments in defense of Atomism (section IV) and second in defense of Isolationism (section V). We conclude that both approaches are unconvincing. However, we will then claim that Caney's most recent proposal for a Holist–Integrationist method of emissions allocations suffers from lack of practicality, whereas EE does not (section VI). Pointing out that, at a closer look, EE is not wholly Isolationist either and would make the world much more just, we argue that – under present circumstances – EE should be favored over Caney's five-step emissions allocation procedure.

IV Holism vs. Atomism

It is usually acknowledged that there are different strategies to respond to anthropogenic climate change. The most prominent and widely discussed strategies are mitigation (defined as either reducing GHG emissions or reducing atmospheric GHG concentrations) and adaptation (adapting to changing climatic conditions). If climatic conditions change and adaptation does not occur or is (partially) unsuccessful, a further possibility is to compensate those harmed by climatic changes (rectification).[8] Obviously, each strategy is associated with costs. When thinking about how to distribute the costs in a just way, one can either consider the costs associated with each strategy separately (Atomism) or think about how to distribute overall climate-induced costs (Holism) (see Caney, 2012). In the past, we have

[7] In sections IV and V, respectively, each distinction will be introduced in greater detail. Caney's two distinctions are based on the assumption that the debate centers on the distribution of overall climate-induced costs. In contrast, the primary aim of EE is to divide a remaining sink capacity. In this chapter, we accept Caney's assumption.

[8] A further possibility might consist in so-called climate engineering (see Rickels et al., 2011).

adopted an (moderately) Atomist approach. We proposed adopting 'Contraction and Convergence' (Meyer, 2000) in combination with a scheme of financing adaptation in which wealthy, high-emitting countries ought to provide resources to the global South (Ott and Baatz, 2012). We thus applied a certain normative principle to the realm of mitigation (equality plus a transition period to account for the status quo) and another principle to the realm of adaptation and rectification (a combination of the polluter pays and the beneficiary pays principles; for detail see the reference). In contrast, proponents of a Holistic position assume that one principle (or a combination of different principles) should apply to all realms of climate change: that is, a single (combination of) principle(s) should govern the *sum* of responsibilities for mitigation, adaptation, and rectification (see Bell, 2008; Caney, 2005, 2009, 2012). Caney (2005, 2006, 2010a), for instance, argues that the overall sum of climate-related costs should be distributed according to a certain combination of the polluter pays and the ability-to-pay principles.[9] Note that the distinction between Atomism and Holism is not categorical but *gradual* in that an approach can be more or less Atomist/Holist, depending on how many realms of climate change are treated separately from each other (as a package).

IV.1 A First Challenge – EE Ignores Important Aspects of Mitigation

EE and policy proposals based on the notion of EE (most notably 'Contraction and Convergence') have been criticized for ignoring important aspects of mitigation. Emissions matter only insofar as they increase the atmospheric concentration of GHG, which ultimately leads to climate change. But another factor substantially influences atmospheric GHG concentrations: GHG sinks. EE, the charge is, ignores sinks and perhaps even all emissions generated by land use and land use changes. "Isolating GHG emissions when our focus should be on anthropogenic forcing as a whole is a mistake" (Caney, 2009: 131). EE is thus criticized for being 'super-Atomistic' in that it does not even consider all relevant aspects of – and costs associated with – mitigation. Ignoring sinks is indeed problematic and *all* factors that have an effect on the atmospheric GHG concentrations must be taken into account. But we claim that EE can do so.

For a start, Caney's claim is somewhat misleading. Instead of emissions, anthropogenic forcing should matter, he claims. However, the things Caney mentions in this respect are natural and artificial GHG sinks. Sinks sequester GHG. This is the reverse process of emitting GHG. Hence, the creation of GHG sinks can be conceived of as negative emissions. Or to put it in more practical terms, if I heat my house (because it is cold outside), cut down some trees in my

[9] Holist approaches are also adopted by Edward Page (2008, 2012) and David Miller (2008).

garden, and plant some new ones afterwards my activities have a net effect in terms of emissions/radiative forcing. While some activities release GHG into the atmosphere, others sequester it. My net emissions budget can be negative, although in this fictitious example it is probably positive. The example highlights that it is not problematic to express what Caney calls anthropogenic forcing in the terminology of (net) emissions. While increasing GHG sinks can be indicated in terms of negative emissions, destroying sinks can be indicated in terms of positive emissions.

Thus, EE can and must take all emission generating as well as emission sequestering activities into account. Obviously this makes it more complex to determine equal shares of the overall emission budget. However, all reasonable approaches must deal with these issues. How to determine *net emissions* caused by complex processes such as land use changes is a general problem of accounting that is not specific to EE or any concept that treats emissions in isolation from other domains of climate change. Therefore, the argument that EE ignores important aspects is wrong – or at least only pertains to specific proposals and thus cannot challenge the notion of EE as such.

IV.2 A Second Challenge – EE Ignores Other Climate-Related Burdens

As stated, EE employs the principle of equality within the realm of mitigation without considering how costs for adaptation and rectification are distributed; it thus treats mitigation separately from the other realms of climate change. According to a Holist position, this is flawed, for the same normative considerations apply to mitigation, adaptation, and rectification and, hence, there is no reason to treat them differently.[10] If so, (P3), which says that "if a global commons is owned by everyone equally, the right to use it should (*pro tanto*) be distributed equally amongst all," is flawed. Rather, one should endorse a principle like the following:

(P3′) If a global commons is owned by everyone equally, the costs related to its (over-) use and maintenance should be distributed equally amongst all.[11]

If a Holist position and (P3′), respectively, are accepted, it seems difficult (even impossible) to defend EE. It is very unlikely that future emissions entitlements ought to be distributed equally when costs for adaptation and rectification are included. First of all, the costs for adaptation will vary considerably around the

[10] Holism does not object to the IPCC's claim that adaptation has specific societal, economic, and moral characteristics, which differ from mitigation (2007: chapter 18).

[11] Note that the egalitarian principle "should be distributed equally" could be a reason to reject (P3′). Holist approaches usually draw on different combinations of polluter pays, ability-to-pay, and beneficiary pays principles (see, e.g., Caney, 2010a; Page, 2008).

world (to put it mildly). Second, while some agents will face compensatory duties to different degrees, others will have no such obligations. It is highly implausible that the entitlements (regarding emissions entitlements) and duties (regarding adaptation and rectification) balance each other in that all agents ought to bear equal burdens in the end (so that emissions entitlements ought to be distributed equally). Rather, 'wealthy polluters' ought to bear much greater climate-related burdens than, say, 'poor non-polluters'. In conclusion, EE can only be defended if mitigation burdens are allocated separately from other climate-related burdens. In order to defend EE, an argument is required for why this should be done. In the following, we will therefore discuss three independent arguments that try to establish that different normative considerations apply regarding mitigation, on the one hand, and other responses to climate change, on the other. That is, these arguments dispute the Holist claim that the same normative considerations apply to all fields of climate justice.

IV.3 Vanderheiden's Arguments for a Separate Treatment of Mitigation

Vanderheiden (2008: 229–30) argues that mitigation is a matter of distributive justice while funding adaptation is a matter of corrective justice (the same claim is made by Risse, 2008: 38).[12] Similar to our approach, he argues for a roughly egalitarian principle regarding mitigation and employs the polluter pays principle regarding adaptation. That is to say, polluters ought to make up for their excessive past emissions by financing adaptation measures as a way of settling their compensatory duties vis-à-vis potential climate change victims; and this is a matter of corrective justice.

We think that Vanderheiden's argument is unconvincing, though. He fails to acknowledge that funding adaptation is *not only* a question of corrective justice. Certainly, corrective justice is of great importance in the realm of adaptation, as we have ourselves argued (Baatz, 2013). However, current polluters that are obliged to finance adaption in order to settle their compensatory duties must not (and cannot) be held accountable for all harm caused by anthropogenic climate change. This would not only overburden them but also be unjust, for they would have to pay for harm caused by others. In order for that harm – that polluters are not responsible for in terms of compensatory justice – to be addressed, many have argued for a hybrid account combining the polluter pays principle and the ability-to-pay and/or beneficiary pays principle. The ability-to-pay principle and some versions of the beneficiary pays principle are principles of distributive (rather than corrective)

[12] Vanderheiden and Risse do not consider rectification.

justice (in detail see ibid.). Thus, funding adaptation is a matter of *both* (re-) distributive and corrective justice.[13] If this is accepted, Vanderheiden's claim that mitigation is a matter of distributive justice while funding adaptation is a matter of corrective justice cannot be sustained and thus cannot justify treating both realms separately.

In a more recent paper, Vanderheiden (2011) also argues that it is wrong to treat mitigation and adaptation as commensurable burdens. If these burdens are commensurable, it does not matter to what extent one engages in mitigation or adaptation activities, as long as one takes over the total costs assigned to her. But why is this a problem? Vanderheiden (2011: 68) offers several reasons. First, from the moral point of view it is better to prevent harmful events in the first place than to try to avoid the harm by successful adaptation. Second, it is usually taken as valid that mitigation is more cost-effective than adaptation in terms of preventing harm.[14] Third, different people will benefit from mitigation and from adaptation activities (Vanderheiden, 2011: 70). Neglecting mitigation burdens, for instance, means that an agent continues contributing to harmful actions. Vanderheiden therefore argues that "allowing states to treat mitigation and adaptation imperatives as commensurable, and to shift resources between the two at will, imposes externality costs when adaptation efforts displace required mitigation actions. For this reason, such cost-shifting should be prohibited, and mitigation and adaptation burdens separately assigned" (Vanderheiden, 2011: 69).

We think that Vanderheiden is absolutely right. This kind of cost shifting, however, does not follow from Caney's argument that the same normative considerations apply to both mitigation and adaptation. Caney simply claims that the costs resulting from the different responses should be distributed according to the same principle(s). This view is silent on the right 'mix' of responses. The Holist claim is that whatever the costs for mitigation and for adaptation will be, they should all be distributed according to the principle X (or X, Y, and Z). Vanderheiden's argument rather is that the right 'mix' of responses matters, that is, it makes a difference whether in a given period ten units of mitigation and five units of adaptation are undertaken, or vice versa. But Holists can consistently endorse this claim. Vanderheiden's argument is concerned with what kind of burdens there are (and how great mitigation and adaptation burdens are) and this is different from how to distribute these burdens. As long as it is safeguarded that enough mitigation is undertaken, mitigation and adaptation do not have to be allocated

[13] Vanderheiden's argument has also been criticized by Caney, who concludes that corrective and distributive justice do not "neatly [map] on to the difference between adaptation and mitigation" (2010b: 35). Note that Caney's argument allowing him to arrive at this conclusion is different from ours.

[14] This holds true at the general/global level. From the perspective of single agents such as states or individuals, adaptation will be more cost-effective if it is assumed that others do not mitigate emissions sufficiently.

separately.[15] In sum, Vanderheiden's argument regarding the incommensurability of mitigation and adaptation burdens is convincing, but it does not show that the holist position is flawed.

In the following, we will consider a rights-based argument that (initially) looks more successful in justifying a separate treatment of mitigation.

IV.4 A Rights-Based Argument for Treating Mitigation Separately?

Different kinds of individual rights exist that are associated with (different) corresponding duties. The argument we will develop in the following starts from the premise that there are two basic duties: the duty to respect the rights of others and the duty to remedy the violation of rights. We conceive of the former as refraining from an action that will violate the right of another person.[16] This also includes refraining from contributing to a harmful action. We conceive of the latter as providing redress to those who suffer from a rights violation.[17] We consider both duties to be widely accepted. What follows from applying these two basic duties to anthropogenic climate change?

Peoples' rights are and will be violated by climatic changes. These changes result from increased atmospheric GHG concentrations that can be reduced by mitigation activities. Harm can thus be prevented and rights respected by engaging in mitigation. More precisely, an agent does not violate rights or refrains from further contributing to rights violations if she sticks to her fair share of entitlements to use the earth sink capacity (however fair shares are defined). Refraining from the violation of rights therefore is about mitigation (see also Shue, 2014: 306–10).

The duty to remedy consists of compensating those suffering from rights violations caused by climate change. That is, the question how to deal with the consequences of *past* overuse of the sink capacity of the earth must be answered. This means to determine a just distribution of remedial responsibilities. The duty to remedy the violation of rights concerns adaptation and rectification. The reason for this is due to a special feature of climate change. Usually, and as stated previously, remedial duties regarding rights violations concern measures to make up for, or at least ease, a suffered loss. The reason is that in these 'usual cases' the harm occurs immediately or shortly after the harmful action. This is different in the case of climate change. The substantial time lag between GHG emissions and negative

[15] On the other hand, this implies that whenever a global institution safeguarding the right 'mix' of responses is lacking, Vanderheiden's argument applies. That is, Vanderheiden's criticism nevertheless applies to all those holist proposals that fail to safeguard the right 'mix'.

[16] We only refer to violations and abstract away impairing and compromising rights.

[17] The remedial duty is formulated so as to allow it to be associated with both positive and negative duties. If the existence of positive duties is denied, a more narrow definition is in order, stating, for example, the duty to compensate those harmed by one's action. We consider the denial of positive duties to be untenable, though.

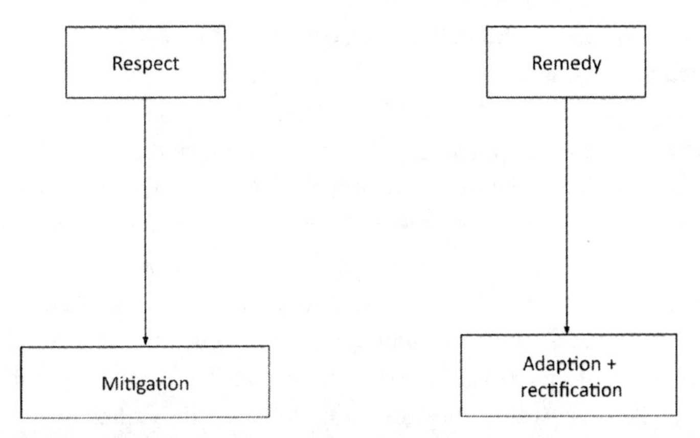

Figure 8.1 Rights are respected by mitigation: that is, an agent does not exceed her fair share of net emissions. The violation of rights is remedied by adaptation and rectification measures.

impacts, usually lamented as a factor that makes it harder for humankind to adopt proper responses (e.g., Gardiner, 2011), is an advantage in this case: it allows for the adoption of protective measures before a potentially harmful event occurs.[18] Hence, an ex ante form of compensation exists as well: to prevent as much harm as possible by financing adaptation projects and programs in regions vulnerable to climate change. This is the reason why remedial duties include both adaptation and rectification.

The argument made so far can be illustrated in the following way:

The proposal is able to distinguish between two general realms of climate justice. One realm is about respecting the rights of others and deals with how to distribute *net* emissions fairly, that is, by determining each agent's fair share of the earth's remaining sink capacity. The other realm concerns how to remedy rights violations due to excessive emissions that already occurred or will occur if no other preventive action is undertaken. The distinction therefore justifies a (moderate) Atomism.

However, the argument only holds if it is not possible to respect the rights of others by means of adaptation/rectification and, moreover, if it is not possible to remedy rights violations by means of mitigation. That is to say, the argument only holds if does not make sense to include arrows from "respect" to "adaptation + rectification" and from "remedy" to "mitigation". Why is that? The argument so far is that respecting rights is *only* about mitigation and that remedying rights violations

[18] This entails several requirements, however: first, that there is sufficient knowledge about future impacts; second, that measures to counter these impacts exist; and, third, that the resources to implement these measures are available. Our claim is that in some (probably even in many) cases conditions one and two are fulfilled and that it is in 'our' responsibility to realize condition three.

is *only* about adaptation/rectification. According to this argument, we are dealing with two separate realms in which different normative considerations apply.

Let us deal with a potential arrow from "respect" to "adaptation + rectification" first. The time lag between potentially harmful actions and actually harmful events generates a substantial challenge to the proposal. As just explained, in principle it is possible to prevent an event from being harmful by adapting ex ante measures that reduce one's vulnerability to such an event. To provide a simple example: although climate change will lead to sea level rise and at some point a grave flood will hit coastal town A, citizens of A and their property might not be harmed because of improved sea walls. If agents C, D, and E are responsible for climate change but pay for improving the sea walls, they do not, in the end, violate the rights of the A citizens. That is to say, they respect the rights of members of A. This means that one can respect the rights of others by financing adaptation as well and that, in turn, the distinction between both duties (to respect and to remedy) does *not* map completely onto the distinction between mitigation and adaptation/rectification.

But we claim that this forceful counter-argument moves too quickly and that it is indeed *not* possible to respect the rights of others *fully* by means of adaptation. Once excessive amounts of GHG are emitted, additional actions are required to prevent rights violations. That is, in the moment of excessive emissions, rights are in limbo. Others are put at risk. If I deliberately risk violating your right to X, am I fully respecting this right? We do not think so, but arguing for this position would take up too much space and we simply sidestep this issue here, for it does not change our conclusion later.

Instead, we will turn to a practical argument for why it is not possible to respect rights fully by means of adaptation. The preceding example of the coastal town is highly stylized – and unrealistic. Most effects of climate change are much more complex than sea level rise, and even adapting to sea level rise is often highly complex. As suggested previously, substantial knowledge about the likely climatic changes and their effects on socio-economic systems is required to adapt successfully. Also, adaptive strategies must exist and must be affordable. While proper and comprehensive adaptation measures are – in principle – able to prevent many rights violations that would otherwise occur, they will not be able to prevent all harms. Some effects will not be anticipated and some losses cannot be prevented. We therefore claim that it is not possible to respect rights fully by means of adaptation because adaptation cannot prevent all rights violations because of barriers in terms of knowledge and resources. But, even if it is not possible to respect rights fully by means of adaptation, this does *not* establish that rights can only be fully respected by mitigation.[19] When atmospheric GHG are already

[19] We are indebted to Pranay Sanklecha for alerting us to this issue.

elevated (but not so high that dangerous climate change is inevitable) it will only be possible to prevent rights violation by a mix of comprehensive mitigation and adaptation policies. In such a situation, respecting rights will require some adaptation in addition. This calls into question the claim that respecting rights is only about mitigation all the time.

Let us now deal with a potential arrow from "remedy" to "mitigation". Is it possible to remedy (possible) rights violations by means of reducing net emissions?

To remedy rights violations via emissions reductions is possible only if an agent reduces emissions beyond what is required in terms respecting the rights of others: that is to say, if she reduces her emissions below her fair share of emissions entitlements. As long as reductions are required to meet the fair share, they count as efforts to respect rights (see preceding discussion). To make up for past excessive emissions in this way is extremely unlikely at present – irrespective of how fair shares are determined in detail and whether we are dealing with countries or individuals.[20]

Perhaps a more realistic case is country A reducing its emissions to its fair share and providing clean technologies to country B in order to help B reduce its emissions.[21] The lower GHG concentrations that result from this will lead to less climate-induced harm. Remedying rights violations is about reducing the harm dimension – ideally to zero. Thus (indirectly) reducing net emissions should count as a remedial action. The fact that I can remedy my own or others' harmful action by deciding not to undertake the harmful action in the future may be a surprise: if I slap someone in the face, I can usually not make up for it by simply refraining from future slaps. Climate change, however, is different in that i) emission generating activities only cause harm when undertaken excessively by a sufficiently large number of agents, and ii) a certain level of emissions is required to live decent lives. From these two well-known facts follows the threshold or fair share conception of illegitimate/harmful emissions (in detail see Baatz, 2014).[22] The fair share concept, in turn, allows direct compensation for the excessive emissions of others: if A (directly or indirectly) reduces her emissions by two units below her fair share, this makes up for the two units by which B exceeds her fair share. Because of the time lag between emissions and

[20] On the level of nation-states, countries usually believed to bear remedial duties are high emitting and/or wealthy. All these countries exceed their fair share by far; however, the remaining small global emissions budget is split exactly. Roughly the same holds true for individuals (see Baatz, 2014), although there might be some exceptions. Poor people who emit less than they are entitled to do not have remedial duties (see Duus-Otterström and Jagers, 2012).

[21] Duus-Otterström and Jagers call this "indirect mitigation" (2012).

[22] The basic thought behind this is that once a total emissions budget is determined that (probably) allows avoiding morally repugnant outcome X, this budget must be somehow allocated between current (and future) human beings. If this allocation is just, everyone receives her fair share of emissions entitlements of the overall budget.

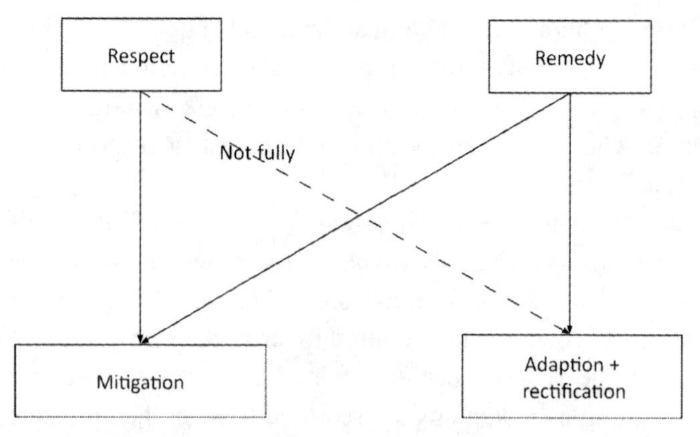

Figure 8.2 Respecting rights and remedying rights violations by means of mitigation and adaptation/rectification.

climate change impacts I can, at least in principle, also compensate for my own excessive past emissions (see Figure 8.2). The upshot is that remedying climate change harm (potentially) caused by excessive GHG emissions is *not only* about adaptation and rectification. Figure 8.1 can thus be supplemented in the following way:

Before moving on, let us pause and summarize the argument set forth in section IV. At the outset we argued that EE can only be defended if mitigation costs are to be distributed independently of costs concerning adaptation and rectification. We then discussed two arguments from Vanderheiden for why mitigation and adaptation costs should be allocated according to different principles. We concluded that both arguments fail. Subsequently, we presented an argument that aims at justifying the separate treatment of mitigation by drawing on the distinction between respecting rights and remedying rights violation. According to this argument, mitigation is only about respecting rights and adaptation/rectification only about remedying rights violations. If so, one can justify that different normative principles apply regarding mitigation (e.g., the principle of equality) and adaptation/rectification (e.g., the Polluter Pays Principle), respectively; that is, how costs for adaptation and rectification are distributed is not relevant for distributing mitigation costs. However, this argument, too, fails because i) respecting rights will require some adaptation in addition to mitigation in certain situations and ii) it is in principle possible to make up for past excessive emissions by reducing current emissions even further. Therefore, normative considerations based on remedial duties can also be relevant in the realm of mitigation and those based on duties to respect in the realm of adaptation. In sum, this argument also fails to establish a principled distinction between mitigation, on the one hand, and adaptation/rectification, on the other.

V Integration vs. Isolation

So far, we have taken for granted that equality is the right pattern of climate justice, that is, climate-related burdens, or burdens associated with one of the realms of climate change, ought to be distributed in an egalitarian manner. Of course, EE can be criticized for employing the wrong pattern of justice and because, say, emissions entitlements should be distributed according to the maximin rule (prioritarianism rather than egalitarianism). Moreover, EE can be criticized for employing the wrong currency of justice because it is concerned with how resources are distributed rather than, say, capabilities. However, there is no need to discuss these claims in greater detail, for they point to a general criticism. The charge is that it is wrong to deal with emissions entitlements in isolation from other considerations of justice.

In this respect, Simon Caney and Derek Bell have argued for an Integrationist approach. They argue that it is a mistake to treat climate responsibilities in isolation from other considerations of global justice such as trade, development, poverty, and health (Bell, 2008; Caney, 2009). One could add migration, education, nature conservation, and the like. Rather, when reflecting on what to do about climate change we should do so in an integrationist fashion: we should treat the ascription of climate responsibilities in *conjunction* with considerations about global justice in general (Bell, 2008: 254; Caney, 2012: 271).[23] The reason is that theories of distributive justice are usually concerned with fair shares of packages of goods and do not deal with single goods in isolation.[24]

EE does exactly this: it distributes one good – emissions entitlements – in isolation from the distribution of other goods. Singling out emissions entitlements must be justified by providing a reason for why EE deserve this special treatment. In section VI we will provide practical reasons for this special treatment. According to Caney, neither do emissions entitlements have the special status of goods that are indeed singled out for separate distribution (as e.g., voting rights), nor can they be associated with specific bundles of goods (Rawlsian primary goods, Dworkinean resources, etc.), nor with Walzer's spheres of justice (2009, 2012). Thus, drawing on established theories of ideal distributive justice it is hard to see why the distribution of emissions entitlements matters in isolation: "it does not make sense to refer to *the* fair distribution of greenhouse gases" (Caney, 2012: 271).

Caney underscores his argument by referring to the well-known fact that, in many cases, wide or even narrow[25] substitutes for GHG emissions exist. That is,

[23] In section VI *one* way of employing a method of Integration is discussed.

[24] The most prominent exception is Michael Walzer's (1983) theory of justice.

[25] "Wide substitutability occurs when one substitutes one kind of good with another quite different kind of good without detriment to that person because their overall share of goods remains just. . . . X and Y are substitutes in

the precise goods that are associated with emissions entitlements can be provided in other ways (by employing renewable energies, adopting energy efficiency measures, changing agricultural practices, changing modes of transport, etc.). In line with his general interest-based theory of justice, what matters to Caney from the moral point of view are the interests that emissions serve. If distributive justice is concerned with the benefits associated with GHG emissions and the interests the benefits serve, respectively, we should focus directly on the fair treatment of those interests, Caney argues (ibid.: 288–90). Thus, it does not make sense to distribute emissions in isolation from other (important) goods (see also Hayward, 2009).

If Bell's and Caney's argument is correct, EE does not even follow from an egalitarian (pattern) resourcist (currency) theory of justice. Recall that (P3) stated that "If A global commons is owned by everyone equally, the right to use it should (*pro tanto*) be distributed equally amongst all". According to an Integrationist position, (P3) has to be rephrased, for example, in the following way:

> (P3″) if a global commons is owned by everyone equally, the rights to use it should (*pro tanto*) be distributed such that the currency of justice is distributed equally amongst all.
>
> *(Seidel, 2013: 4)*

That is to say, the distribution of emissions entitlements is supposed to level unequal endowments with Rawlsian primary resources, capabilities, or whatever metric is used as the currency of justice. Because of the extremely disparate endowments with capabilities, resources, freedoms, etc., of the people around the world, distributing emissions entitlements according to (P3″) will not result in an equal distribution of emissions entitlements. Also, it is extremely unlikely that employing any other distributive pattern (prioritarianism, sufficientarianism) would 'accidentally' result in an equal distribution of emissions entitlements. The upshot is that if Integrationism is correct (and Isolationism wrong), EE seems highly unconvincing.

We consider the case for an Integrationist approach to be a strong argument against EE. In the following we will critically investigate what can be said in defense of both Integrationism and Isolationism and, hence, EE.

V.1 A Libertarian Argument in Defense of Emissions Egalitarianism?

Moellendorf (2011) discusses different ways to justify EE. After arguing that, prima facie, right libertarians, left libertarians, and egalitarian liberals should endorse EE, he engages with Caney's criticism of EE and concludes that Caney provides strong reasons at least for liberal egalitarians to abandon EE. Regarding

the narrow sense when X and Y both possess the same kind of properties: for example, they produce the same specific kind of benefit (and have the same kinds of disadvantages) and thus can be used interchangeably to achieve that benefit" (Caney, 2012: 283–4).

libertarians he makes the interesting suggestion that "many egalitarian liberals will find the objection pressed by Caney relevant, but it is not likely to disturb libertarians who think that the justice of distribution depends on whether they respect property rights, not whether they meet needs" (Moellendorf, 2011: 117). That is to say, if rights trump needs and if EE is based on common ownership of the earth's sink capacity that implies property rights, Caney's criticism may lose its force. The reason is that to libertarians respecting rights matters more than meeting needs and, therefore, it is (*pro tanto*) unjust to violate a person's legitimate right in order to fulfill another person's needs.

If so, it makes sense to treat emissions in isolation from other considerations of global justice, for example, poverty.[26] If property rights matter more than needs, benefits from using my part of the earth's sink capacity belong to me and are not supposed to make up for the lack of resources of others, or so one might argue along the lines sketched by Moellendorf.

To show that libertarians should be committed to EE, Moellendorf argues that as a result of Locke's famous proviso, individuals are not free to appropriate as much of the earth's sink capacity as they can (by emitting great quantities of GHG). Rather, total use of the earth's sinks must be limited in order to leave "enough, and as good . . . for others" (Moellendorf, 2011: 106). It follows that a cap on total emissions must be introduced. "If a person emits more than the amount that every other can emit without climate perturbations, the proviso has been violated since this leaves smaller shares for everyone else unless climatic perturbations are to occur. According to these Lockean accounts, emissions beyond one's share of the threshold violate the natural property rights of humanity, whose members own the atmosphere in common" (Moellendorf, 2011: 107).

From the claim that the sink capacity is commonly owned, Moellendorf (2011: 108) jumps to the conclusion that "each person has an equal share of the earth's atmosphere". The conclusion gains plausibility if it is based on the assumption that the sink capacity is *equally* owned by all. That is to say, an argument is required for why (right) libertarians, who are usually not highly concerned with egalitarian distributions, should endorse (P2).[27] In this case, one might argue that since criteria of just appropriation of resources through one's own labour do not apply, distributing ownership rights equally remains the only option.

Now, egalitarian ownership can be conceived of in different ways. According to the equal division view, equal ownership of the sink capacity implies owing equal *parts* of the sink capacity (Risse, 2008: 14–15). Risse argues that conceiving of

[26] Caney discusses further potential defenses of Isolationism. We agree that most of these cannot justify this approach regarding emissions entitlements (in detail see 2012, 272–82).

[27] As a reminder: (P2) A global commons is owned by everyone equally. Moreover, one could argue that each individual would owe an equivalent share of the totality of all global commons, rather than of this specific commons. For the reasons provided in section III we do not discuss challenges to (P2).

equal ownership as equal division leads to implausible consequences; collective ownership of the earth and/or its resources does *not* imply owning any particular object or part of it (Risse, 2008: 27–28). If this were the case, each individual would owe an equal part of, say, the oil in the ground of Saudi Arabia. This seems counter-intuitive. Risse argues in favor of a different conception of egalitarian ownership according to which "all co-owners ought to have an equal opportunity to satisfy their needs to the extent that this turns on obtaining collectively owned resources" (Risse, 2008: 13). This understanding of egalitarian ownership, however, implies (P3′) rather than (P3)[28] and therefore cannot justify Isolationism.

Perhaps, however, the earth's sink capacity is a specific case of common ownership that is different from common ownership of 'particularly located' resources such as land, oil, or minerals. Because of this difference, the argument could continue, common ownership of earth's sink capacity should be conceived of according to the equal division view. However, such an argument is yet to be made.

Moreover, Moellendorf's sketch of a libertarian argument in favor of EE faces a second challenge. To us, it is not clear why ownership rights of the earth's sink capacity trump considerations regarding (basic) needs. The property rights that (right) libertarians are so keen to protect from infringements are different from the right to own parts of the earth's sink capacity. As pointed out by Moellendorf himself and others, the sink capacity is not created through investing one's labor, time, and/or money. The argument that there is no duty to surrender my property to fulfill the needs of others gains force by the assumption that my investments have increased the value of a given natural resource (e.g., converting some piece of land into farmland) and that, perhaps, others even benefit from my converting these resources (they can buy the crops I grow). But this is not the case regarding the earth's sink capacity; there is no just appropriation *sensu* Locke. Hence, it is far from clear that ownership rights of the sink capacity have the high status (right) libertarians usually ascribe to property rights. But only if they have this status does it make sense to treat emissions entitlements in isolation from considerations regarding the fulfilment of (basic) needs.

These few critical remarks are not able to show that a libertarian argument for EE cannot be made. Still, Moellendorf's argument would need more elaboration, and in the present form it is not able to defend Isolationism. In the following, we will develop an argument that further undermines the case for Isolationism (and thus strengthens Integrationism).

[28] While (P3) says that "if a global commons is owned by everyone equally, the right to use it should (*pro tanto*) be distributed equally amongst all", (P3′) reads, "If a global commons is owned by everyone equally, the rights to use it should (*pro tanto*) be distributed such that the currency of justice is distributed equally amongst all".

V.2 An Additional Argument in Favor of Integration: Financing Adaptation

In the academic debate, as well as in this chapter, the discussion regarding Integration vs. Isolation focused on mitigation and the distribution of GHG, respectively. However, other realms of climate change can be treated in an Integrationist or in an Isolationist fashion as well. Think of adaptation. The IPCC (2007: 869) defines adaptation as an "adjustment in natural or human systems in response to actual or expected climatic stimuli or their effects, which moderates harm or exploits beneficial opportunities". Although definitions of adaptation differ, they usually refer to the avoidance of climate-induced harms. In a review on financing adaptation Nicole Hartzell-Nichols (2011: 688) writes that "generally, adaptive measures are meant to reduce the harmfulness of climate impacts (climatic changes)". She later continues, "How adaptation is understood depends on who or what we value and who or what is adapting" (2011: 689). In particular, it depends on what is considered to be harmful, that is, how harm is defined.

The standard definition of harm is counterfactual: A is harmed by X, if X makes A worse off than A would otherwise be (i.e., without the occurrence/action of X). In the intergenerational context, in which climate is situated, the counterfactual conception is vulnerable to Parfit's Non-Identity Problem (1987: 351–79). According to the Non-Identity Problem, most people will *not* be made worse off by climate change than they would be otherwise because climate change is, at the same time, a necessary condition for their very existence.[29] This consequence can be avoided by employing a comparative or an absolute conception of harm. According to the comparative conception A is harmed by X, if X makes A worse off than others are at present. This understanding of harm, however, does not capture well what is meant by adaptation; it would mean that one is harmed by climate impacts if one is made worse off than others are at present. Instead, the absolute (or threshold) conception saying that 'A is harmed by X, if X deprives A of what A is entitled to' is better suited in the context of adaptation. According to this definition, all those are harmed by climate changes that are pushed below a certain threshold by climatic changes. A definition of adaptation based on this understanding of harm reads as follows: "Designing natural and social arrangements so that people are able to cope with climate-related threats and exercise their legitimate entitlements without loss" (Caney, 2009: 127). This is an Integrationist definition because a reference to general entitlements is made and what these

[29] At the heart of the Non-Identity Problem is the person-affecting view according to which a course of action can be morally wrong only if a particular person will be harmed, that is, made worse off. Moreover, the course of action in the past and present determines which particular individuals will be born in the future. Thus, depending on the course of history, different individuals are born. In a counterfactual world (e.g., one without climate change) other individuals would live. Without climate change I would not exist and, therefore, I cannot claim that I would be better off without climate change (we discuss the possibility that one might benefit from climate change in Baatz, 2013).

entitlements are must be determined by a theory of justice. Again, what one is entitled to in terms of adaptation depends on her overall endowment of, for example, resources or capabilities.

An Isolationist definition of adaptation must avoid any reference to entitlements or other general considerations of justice such as well-being or rights. An easy and straightforward way of doing so is to say that 'adaptation aims at avoiding that A is made worse-off by climate change'. Unfortunately, this is no solution given Parfit's paradox, for without climate change A would not exist and therefore it does not make sense to claim that A is made worse off by climate change, as highlighted by the Non-Identity Problem (in detail see Meyer, 2004). But, perhaps, the vulnerability to the Non-Identity Problem can be circumvented by referring to specific (weather) events rather than climate change as such. Then one can, for instance, define adaptation as 'avoiding that A is made worse off by a single event caused by anthropogenic climate change'. That is to say, if A is worse off in t_2 compared to t_1 because of event X, adaptation aims at restoring A to her condition in t_1 or rather safeguards that A stays in that condition in the first place. This definition is not vulnerable to the Non-Identity Problem because the harmful single event is not, at the same time, the reason for A's existence. But this definition causes more difficulties than it resolves.

First, in many cases we do not know whether and to what extent a single event is caused by anthropogenic climate change. Second, single events that worsen conditions of individuals often are hard to determine or do not exist. Many negative impacts of climate change are long term and pervasive. Rather than single events, climate changes leads to gradual changes in climatic conditions, ecosystems, natural processes, etc., and puts pressure on social and economic systems. Also, climate change is often just one of many causes of harm or causes harm only in combination with several other incidents and factors. In consequence, in many cases it will be difficult or even impossible to identify a single harmful event caused by anthropogenic climate change. Third, focusing on single events, even if they can be identified, risks adopting wrong strategies by neglecting the 'big picture'. Consider the following example: a region is getting drier, as becomes manifest in several drought events. Adaptation according to the 'single event definition' would mean to counter each drought by specific measures. Still, approaching the issue from the perspective of a region that will become drier in the long run (e.g., over the next hundred years) might generate more comprehensive and long-lasting solutions than focusing on each specific drought. The upshot is that it is far from clear what the condition of A in t_1 is that is supposed to be sustained *if no reference to general considerations of well-being or entitlements is made.*

Even if the Non-Identity Problem and all of the preceding conditions are not considered, an Isolationist approach to adaptation faces a further drawback.

In many cases it will not be possible to sustain or restore a given status quo (A in t_1). Think of foreseeable climate-induced impacts on health in developing country B. A reasonable long-term adaptive strategy is to improve the health care system of B. But this will, in all likelihood, not sustain the status quo. On the one hand, improving the health care system will also benefit those not actually or only little affected by climate change. On the other hand, although an improved health care system will *ex hypothesi* be able to improve conditions of those negatively affected by climate change, these individuals might still be worse off compared to a counterfactual situation without negative climatic impacts. Thus, while some citizens will still be worse off than in the counterfactual situation, others will be better off. Now, according to the entitlement definition of adaptation,[30] the adaptive strategy in this stylized example is successful for it is able to avoid most suffering *ex hypothesi* (most people are not pushed below the threshold). But, according to the counterfactual conception (adaptation aims at avoiding that A is made worse off by climate change), the same strategy is much less successful, for it is not able to sustain the status quo. This latter assessment, however, seems counter-intuitive.

If these arguments are correct, we have presented a robust case in favor of an Integrationist understanding of adaptation.

V.3 Intermediate Conclusion

Reasons so far put forward in defense of both Atomism and Isolationism are either unconvincing or, at best, highly contested.[31] If this is correct, (P3) should be rejected. The preceding argument cannot justify EE.[32] That is to say, EE fails as an ideal principle. Ideal principles determine what "ought ideally to occur from a moral point of view" (Shue, 2014: 334). Ideals can be realized only by way of approximation. Approximation to an ideal situation can be reached by strategies that Henry Shue has dubbed "extrication ethics": "Philosophers and political theorists have written little about what might be called principles for transitions. ... Normally, we are offered an ultimate ideal and, in effect, wished good luck in figuring out how to reach it. I think this is lazy and irresponsible, but transitions are hard, intellectually as practically. ... I think that extrication ethics largely constitutes transition ethics" (Shue, 2014: 129).

[30] "Designing natural and social arrangements so that people are able to cope with climate-related threats and exercise their legitimate entitlements without loss" (see earlier discussion).

[31] Isolationism abstracts away non-climate-change-related human suffering. Isolationists would have to show under which conditions such abstractions are morally permissible. This would require a closer look at the overall ethical design of (ideal) theories of global justice. Such debate is beyond the scope of this article.

[32] Many thanks to Christian Seidel for suggesting to state this clearly.

Approximation can be either more direct (*intentione recta*) or more indirect (*intentione obliqua*). On a direct route, one tries to specify and apply ideal principles under real world conditions. Caney (2012) made a proposal for how to distribute emissions entitlements under real world conditions starting from an ideal (Holist–)Integrationist position. On an indirect route, one tries to move towards the ideal by using different concepts or principles. If EE fails as an ideal principle, it might still be possible to defend EE on practical grounds as part of a more indirect approximation.[33] We think that extrication ethics is possible on both routes. On both routes, however, practical considerations must be accounted for and EE looks much more promising when compared to real world procedures that include practical considerations. The last section will therefore compare EE with Caney's most recent proposal for a modestly Integrationist method of emissions allocation.

VI The Virtue of Simplicity: A Practical Argument in Favor of EE?

Before actually comparing both approaches it is necessary to introduce Caney's proposal briefly.

VI.1 Caney's Five-Step Procedure

A first argument regarding practicality simply denies that an Integrationist approach is possible at present: COP negotiators have no choice but to decide on how emissions entitlements are distributed in isolation. It is certainly true that emissions must be distributed while other goods are not up for (re-)distribution. But when deciding on the distribution of emissions entitlements endowments with other goods can be *considered* (see Caney, 2009, 2012).[34] This reply, however, provokes a further challenge: distributing emissions so as to make up for all other unjust circumstances will overburden COP negotiators and lead to deadlock (see Meyer and Roser, 2006: 239). In reply, Caney (2012: 278, 292) proposes a moderate Integrationist approach: instead of a maximal account of justice, COP negotiators can endorse a minimal account, which only seeks to identify what is absolutely essential. That is to say, only those things are considered that are relevant for the fulfillment of, say, basic needs. Thereby, it is possible to sidestep disagreement about maximal accounts of justice and political opposition that the

[33] A similar approach is endorsed by Rudolf Schuessler (2014), who defends EE in a "second-order perspective" via focal choice.

[34] Indeed, such a procedure is demanded by (P3′), which states that 'if a global commons is owned by everyone equally, the rights to use it should (*pro tanto*) be distributed such that the currency of justice is distributed equally amongst all'.

employment of a maximal account, such as global equality of opportunity or a global difference principle, would face (Caney, 2012: 278).

Thus, to preclude the criticism that a wholly Integrationist approach (taking *all* considerations into account that matter regarding global justice) is completely unworkable under real world constraints, Caney proposes a moderately Integrationist approach. In combination with Caney's Holist position, (P3) can be expressed as follows (*Holist–Integrationist minimal-justice view*):

(P3‴) If a global commons is owned by everyone equally, the costs related to its use and maintenance should be distributed so as to meet the basic needs of all current and future persons.

How would an allocation according to (P3‴) within a global climate treaty look? Caney proposes a five-step procedure to allocate emissions according to an Integrationist minimal justice position.[35] To this, one would have to add the Holist element, that is, say something on how costs for adaptation (and rectification) are to be taken into account. Given that this would probably make the procedure more complex but certainly not less complex, our arguments that follow concerning practicality will not be undermined by adding the Holist element. Caney's proposal is as follows:

In step 1 (The Normative Starting Point) one adopts a set of principles of intra- and intergenerational justice that determines what people are entitled to as a matter of justice (according to a minimal account). In step 2 (The Sustainability Condition) "one must assess whether the account of distributive justice affirmed in Step 1 makes demands on the natural world that can in fact be met" (Caney, 2012: 293). Step 2 is required by the interdependence between an account of justice and the environment: on the one hand, the realization of any account of justice requires using natural resources and might create environmental hazards; on the other hand, the realization of principles of justice requires that certain natural resources are available and that certain hazards are absent (in detail see Caney, 2012: 293–95).[36] Step 3 (The GHG Implication) determines what step 1 and 2 imply for the distribution of emissions entitlements: how to distribute them in a just and sustainable society? This can be achieved by looking at people's entitlements in different sectors (e.g., food, health, housing, and education) and determining the amounts of GHG required to fulfill these: "By identifying persons' entitlements and then their implications for the use of GHG emissions one can then derive an account of how GHG emissions should be distributed" (Caney, 2012: 297).

[35] This is just one way of realizing a method of Integration. Elsewhere, Caney suggests a scheme in which emissions entitlements are allocated via a global auction, as proposed by Tickell (2008) and Barnes et al. (2009). The scheme faces none of the difficulties discussed later (see Baatz, 2013) but belongs in the realm of political utopias at present.

[36] In her Ph.D. thesis, Lieske Voget-Kleschin (2013) discusses the interdependence between an account of justice and the environment at length; see also Voget-Kleschin (2013, 2015).

Depending on circumstances, entitlements can be met in other ways if (narrow) substitutes for GHG emissions exist, as noted previously. Therefore, realizing people's entitlements may be compatible with a variety of different distributions of GHG emissions (Caney, 2012: 298). This is step 4 (The (Narrow Substitutability Proviso). Because of this indeterminacy a fifth step is required (Indeterminacy and the Role of Institutions). Within a political process at different levels "participants can select from the various different combinations available and agree among themselves precisely how people's entitlements are secured. The only way to move from the multiple possible combinations that are thrown up by Step 4 to any specific outcomes is to have political processes in which the relevant parties decide what particular combination of natural resources will be employed in order to realize people's entitlements (as determined by Steps 1 and 2)" (Caney, 2012: 298).

VI.2 Comparing Emissions Egalitarianism with the Five-Step Procedure

Ethics is supposed to guide actions as well as societal praxis.[37] Therefore, issues concerning practicality/feasibility are an indispensable part of ethical all-things-considered judgements; they are not just an add-on to an already completed ethical analysis. This is even more important when dealing with transition or extrication ethics as introduced previously. In consequence, besides its justness the feasibility of a procedure matters. However, feasibility is a complicated and thorny issue that is often dealt with in problematic ways. To avoid at least some of these problems, we distinguish between institutional and political feasibility. Institutional feasibility is concerned with how well a proposal corresponds to existing institutions and how easily it can be implemented given current institutions. Political feasibility refers to the political will to implement a proposal. Here questions such as 'How likely is it that political decision makers will agree on proposal X?' are relevant. Needless to say, from the moral point of view the proposal should be just. If one endorses the minimal justice view employed previously, a situation is just in which all human beings are able to fulfill their basic needs.

Justice, institutional, and political feasibility are three incommensurable categories. For instance, higher political feasibility does not make a proposal more just and the other way around.[38] Incommensurability implies that the 'ranking' in one category cannot be set off against the other categories (e.g., saying that a score of 10 regarding justice and 5 regarding feasibility yields an overall score of 7.5 is not possible). Rather, a proposal should fare well in each category; and if its

[37] This comparison would warrant a paper on its own. Our aim here is to sketch roughly how EE could be defended.

[38] We are currently working on a more elaborated set of criteria.

assessment in one category is very bad it should not be pursued (a very unjust or institutionally/politically hardly feasible proposal is not worth pursuing – as long as this assessment holds true). Obviously, this will only allow for a rough comparison. But anything more detailed would not make much sense given the comparatively indeterminate criteria, particularly concerning political feasibility. We will now discuss how EE and the five-step procedure fare in each category.

Justice

To cut a long story short: if one thinks that climate change is a matter of global justice, and if one thinks that established theories of (global) distributive justice matter in order to determine what people around the world are entitled to, the five-step procedure is more just than EE. We think that this follows from our previous discussion – although much more could be said on this.

It matters for the comparison, though, whether or not the five-step procedure is much more just than EE. We claim that it is not, or rather, we disagree that EE results in a "possibly very unjust" allocation (as one reviewer put it).

First of all, an equal distribution of emissions entitlements might actually be sufficient to fulfill basic needs of all or almost all human beings. Whether or not this is the case depends on how big the overall carbon budget and on how minimal the account of justice is. The more ambitious the account and the smaller the budget the more unlikely it is that EE will meet the demands of (minimal) justice.

In this respect it should be noted that the substitutability of emissions generating activities highlighted by Caney in order to reject an Isolationist view also is a reason for why EE might not be unjust at all. Representatives of some developing countries are keen to point out that they have a right to development, as recorded, amongst others, in the UNFCCC. And EE would probably not allow them to go for industrial development as currently taking place in China. But emissions are just one means amongst others to reach development goals. In principle, these can be reached without generating high emissions. The fulfillment of basic needs can be accomplished by employing renewable energies, energy efficiency measures, and the like. Of course, developing countries would need more technical and financial assistance by developed countries to realize such developmental trajectories. But it is not true that EE as such does not allow for fulfilling (basic) needs of the citizens of developing countries.

Moreover, the extent to which development necessarily results in high emissions also depends on one's understanding of development. As Amartya Sen (2001) and Martha Nussbaum (2007) have argued for decades, what matters is not economic growth but the freedom to realize certain functionings. This is not necessarily linked to high emissions. By way of example, in 2011 the United Nations Developmental Program (UNDP) published a report that investigated the relation between carbon dioxide emissions per capita and the Human Development Index (HDI), a composite index combining indicators of life expectancy, educational attainment, and income. The report concluded that the association between carbon dioxide emissions per capita and the components of the HDI "is positive and strong for income, still positive but weaker for the HDI and nonexistent for health and education" (UNDP, 2011: 25). The important upshot is that EE is not *necessarily* in conflict with the modest Integrationism for which Caney argues.

Finally, EE in itself would indeed be unjust in that it ignores past emissions and climate change impacts, respectively. However, we have argued that EE (in the form of Contraction and Convergence) ought to be combined with a scheme of financing adaptation that comprises a polluter pays and a beneficiary pays/ability-to-pay component (Ott and Baatz, 2012; see section IV). This renders EE much more just overall.[39] If EE plus such an adaption scheme would be implemented, this would be a huge improvement compared to the status quo and would make the world much more just. And EE would certainly improve the situation of the poor. As said, the five-step procedure would result in an even more just distribution of emissions entitlements. However, compared to the current extremely unjust distribution of goods as well as de facto emission entitlements the difference seems rather small.

Institutional Feasibility

Compared to EE, the five-step procedure is extremely complex. We will first highlight the complexity by mentioning three aspects and then say why we think that this is a problem.

First, as explicitly acknowledged by Caney, if the account of justice (step 1) violates the sustainability condition (step 2), the account must be revised until step 1 and 2 are compatible. For instance, the amount of emissions entitlements that is available today must be limited in order not to violate other entitlements of future generations. However, how much of emissions entitlements are actually needed to fulfill current individuals' other entitlements becomes clear only after step 3 is accomplished. It may turn out that fulfilling the entitlements established in step 1 requires more emissions than available according to the sustainability condition (step 2). In this case, the account of justice must be revised and the whole process (steps 1–3) must be repeated.

Second, the five-step procedure starts as a derivative approach but ends by pointing to "decisions of relevant parties," as stakeholder approaches do. It will be

[39] Note that this is not an ad hoc supplement but rather points out that EE must *and consistently can* be combined with a scheme that deals with adaptation and rectification. Note also that such a scheme must be added to the five-step procedure as well or costs for adaptation and rectification must somehow be integrated in the procedure (see earlier discussion). We are thus not comparing apples and oranges.

a highly contestable issue how much emissions entitlements are required to meet (basic) needs in different circumstances. Emissions entitlements cannot be derived in any logical sense but require value judgements regarding the amount of emissions needed for "decent" housing, "necessary" transport, "appropriate" health care, etc. How many emissions are required also depends on the availability of substitutes (see earlier discussion). To what extent substitutes exist and how (narrow) these are add other contestable issues. Furthermore, all these issues must be discussed and agreed upon in an inclusive way. And in some cases it might even be impossible to agree on a principled and clear allocation of emissions entitlements.[40]

Third, these political processes "at different levels of governance" face a trade-off. The more local the processes are, the better one can decide on how best to fulfill people's entitlements. Assuming that every regional unit wants more rather than fewer emissions entitlements, allocating too many emissions (i.e., exceeding the overall emissions budget) is a likely outcome of many local processes. This can be prevented by determining at the beginning which regional unit gets how many emissions entitlements. But deciding globally on how many emissions are needed in each region will involve substantial generalizations. Thus, while top-down approaches must rely on quite general assumptions, bottom-up approaches risk allocating too many emissions. A solution for this trade-off might be a mix of top-town and bottom-up approaches that are harmonized to the extent possible. Again, this will be a rather complex endeavor.

The five-step procedure is in need of huge information input, particularly regarding steps 3 and 4. This in turn requires transparent data generation and collection, but the institutional capacities to do so are currently lacking in many regions. Also, we doubt that the procedure yields acceptable outcomes in all those regions lacking accountable and democratically legitimized institutions. On the other hand, complexity might not be a problem for the reiterative process of step 1–3, as it "is an intellectual exercise, looking for reflective equilibrium. ... That this system has feedback loops ... seems to be an advantage" (anonymous reviewer's comment). But Caney's proposal is certainly not just an intellectual exercise but also a real world allocation procedure. It must come to an end within a reasonable time frame. If undertaking steps 1–3 takes a couple of years (which we think is an optimistic assumption), it is somewhat problematic to repeat this process two to four times given the urgency of the topic. More importantly, political representatives and citizens must be able to comprehend and perhaps

[40] Think of mobility needs in the United States: to what extent is it justified to visit close relatives who live on the opposite coast by plane? To what extent is one allowed to use the car in a remote rural area? And although individuals only have very limited possibilities to influence the mobility structures surrounding them, the federal and state governments can be blamed to a certain extent for this. How many emissions entitlements do inhabitants of a remote village receive?

conduct the procedure themselves in order to make reasonable decisions. But the technological, political, and philosophical problems involved are huge. Caney's method introduces highly contested and contestable issues (reasonable disagreement) into climate negotiations. Such disagreement is convenient to governments who oppose a robust mitigation regime. Moreover, the extraordinary complexity of these issues would overburden current COP negotiations and probably the political processes mentioned in step 4.[41]

We do not see that the current global governance system is capable of implementing Caney's procedure in an adequate manner. Compare this with tax systems: we believe that a general income tax of, say, 20 percent would be less just than the current complex German progressive tax system that tries to account for real world differences between individuals. To take all these differences, special cases, and exemptions into account requires a rather complex system of accounting – and a huge bureaucracy. If proper institutions are missing, complexity *is* a problem. EE faces none of these problems because of its simplicity.[42] Our general point is that the epistemic demands made by an Integrationist method are much higher and more difficult to fulfill at present.[43]

Political Feasibility

So far, Caney has not commented on the issue of complexity. He has, though, claimed that an additional advantage of this approach is its greater flexibility *and* political feasibility: "Since meeting people's . . . needs can be done in more than one way, it offers those who are responsible for providing energy (and food) more freedom than do alternative approaches such as the equal per capita approach. Second, . . . it may, for example, allow some to have higher emissions than would otherwise be the case, if they generate high amounts of clean technology for others and enable others to boost their energy efficiency. This renders it more politically realistic" (Caney, 2012: 298).

We disagree. We think that EE is more flexible than Caney's statement suggests and that the flexibility of Caney's approach is a drawback rather than an advantage.

[41] We acknowledge that Caney's proposal actually corresponds to political realities: at COP negotiations a method of increasing Integration can be witnessed as more and more issues are dealt with that a global climate treaty is supposed to govern. In consequence, negotiations are becoming ever more complex. We are under the impression that the increasing complexity is one additional reason for the current COP failure.

[42] We conceive of EE as splitting the total emissions budget equally amongst all living human beings. If the budget shrinks, equal shares will be smaller as well (the time-slice view). We do not endorse a 'whole lives view' according to which emissions entitlements are to be equalized over a person's lifetime, for this would add substantial complexity to EE. Thanks to an anonymous reviewer for requesting clarity on this point.

[43] "The whole point of integrationism is that A's relatively (i.e., as compared to B) greater endowment with one good (e.g., emissions) may be justified if A has comparatively (i.e., as compared to B, C, or D) less of other goods (e.g., money) – such that A has a fair share of good G if and only if the overall distribution of all relevant goods across all persons is fair. So, to determine a person's fair share in an integrationist fashion we have to . . . include information about the overall endowments of (all) other persons with (all) other goods" (Seidel, 2014: 40).

On the alleged rigidity of EE: First, EE makes most sense to us if implemented as part of a cap and trade system; that is to say, EE is the initial allocation within a system of tradable emissions entitlements. Allowing for emissions trading would increase the flexibility of EE and would generate income for poor persons with low emissions. Surely, a global trade system is still a long way off. At least in the EU a trade system already exists. Our point is that the more EE is combined with emissions trading, the less rigid it is and the more Integrationist it becomes. Second, EE can account for the aspects raised by Caney. Rather than prescribing a rigid allocation rule, EE may function as a yardstick that allows for deviations. If country A provides renewable energies to country B or directly pays for mitigation measures in B, this can be taken into account by granting A more emissions entitlements than would correspond to A's equal share. Note that this kind of deviation refers to accounting issues that pertain to all approaches (as discussed in section III); they are thus different from claims for more emissions entitlements based on 'special needs' that would open Pandora's box. Also, one can deviate from an equal distribution as a concession to political realities. For instance, COP negotiators might agree on giving country C and D more than their equal share in order to gain their consent. This, however, would not mean to abandon EE as a yardstick.

On the flexibility of the five-step procedure: Caney is right that his procedure leaves more room for negotiation amongst COP parties compared to EE. We consider this to be a drawback rather than an advantage, though. One problem is that the flexibility introduces a great deal of complexity, as argued previously. In the earlier paragraph we argued that EE can function as a yardstick that might allow for deviations. Now, the other problem of the five-step procedure is that the *yardstick itself is up for negotiation* (and bargaining): what people are legitimately entitled to in terms of emissions must be determined in political processes. The less fair these processes are, the more successful unjustified claims for legitimate entitlements will be. And the more complex and contested an issue is, the easier it is to pass off unjust positions as just. To us, this kind of indeterminacy is a clear drawback with respect to COP negotiations. The high flexibility opens the door to arbitrariness under current real world conditions. That is to say, increased political feasibility may occur at the price of a (substantially) less just outcome.

Another point concerning political feasibility should be mentioned. If the five-step procedure is taken seriously, wealthy high emitters will receive even fewer emission entitlements than in the case of EE. The core idea of Integrationism is to take overall endowments into account. Those endowed with many other goods will get fewer emissions in most cases (for they have other means to fulfill their basic needs). Thus, the procedure will result in an even more radical redistribution of emissions entitlements compared to EE. Wealthy high emitters will resist this even more fiercely than they currently resist anything tending in the direction of EE. An egalitarian distribution is a focal point in negotiations in that (poor) low emitters would substantially gain from EE while this is something (rich) high emitters must minimally concede (see also Schuessler, 2014). If EE seems politically unrealistic at present, the political feasibility of the five-step procedure is even lower.

Summary

The five-step procedure would make the world even more just than EE, but it is less politically feasible. If the feasibility of the former were increased by opening up the determination of basic needs to real world bargaining processes, the outcome would be less just. Under current circumstances the institutional feasibility of the procedure is very low. Institutional capacities for its proper implementation are lacking and it would fall prey to current power asymmetries and non-transparent institutions. We will outline in the Conclusion what this means for EE and the five-step procedure, respectively.

VII Conclusion

We have argued that the global commons argument for EE fails. If one wants to defend EE, one must do so on practical grounds. We have hinted at how this can be done. Compared to Caney's five-step proposal EE is less just but marked by higher political feasibility and much higher institutional feasibility. If facing the choice between EE and the five-step procedure, our recommendation to political decision makers is to aim for EE plus an adaptation scheme.

Several points must be noted, however. The arguments presented in section IV are supposed to start a debate rather than offering conclusive proof for our position. Moreover, both our aim and our results are limited in that we compared EE to just one other proposal (although there are not many proposals out there at present). Apart from criticizing our arguments one could also propose a simpler Integrationist procedure. Last but not least, part of our conclusion is that the five-step procedure, if followed as envisioned by Caney, would result in a more just distribution of emissions entitlements. Feasibility is not a static or given phenomenon but can be influenced. Politicians can be voted out of office (in some countries) and institutions can be reformed. We conceive of EE as a first step towards a more just world and a global redistribution of entitlements that provides the poor and voiceless with more (bargaining) power. EE can function as capacity building for more complex allocation procedures. It can help in striving towards more legitimate and capable regional as well as global institutions as part of an "extrication ethics".

Acknowledgements

We are highly indebted to Christian Seidel, who shared his (unpublished) paper draft with us and provided helpful feedback on numerous points. Also, we are very thankful to two anonymous reviewers who urged us to be much clearer throughout the chapter on what our arguments establish and on how we proceed. We acknowledge funding from the German DFG's Priority Program 'Climate Engineering: Risks, Challenges, Opportunities?' (SPP 1689).

References

Agarwal, A., and Narain, S. (1991). *Global Warming in an Unequal World: A Case of Environmental Colonialism*. New Delhi.

Baatz, C. (2014). Climate Change and Individual Duties to Reduce GHG emissions. *Ethics, Policy & Environment*, **17**(1), 1–19.

Baatz, C. (2013). Responsibility for the Past? Some Thoughts on Compensating Those Vulnerable to Climate Change in Developing Countries. *Ethics, Policy & Environment*, **16**(1), 94–110.

Baer, P., Athanasiou, T., and Kartha, S. (2007). *The Right to Development in a Climate Constrained World: The Greenhouse Development Rights Framework*. Berlin: Heinrich Böll Foundation.

Barnes, P., Costanza, R., Hawken, P., Orr, D., Ostrom, E., Umana, A., and Young, O. (2009). Creating an Earth Atmospheric Trust Fund. *Science*, **319**(5864), 724.

Bell, D. R. (2008). Carbon Justice? The Case against a Universal Right to Equal Carbon Emissions. In *Seeking Environmental Justice*. Amsterdam and New York: Rodopi, pp. 239–57.

Bovens, L. (2011). A Lockean Defense of Grandfathering Emission Rights. In: *The Ethics of Global Climate Change*, ed. D. Arnold. Cambridge and New York: Cambridge University Press, pp. 124–44.

Caney, S. (2005). Cosmopolitan Justice, Responsibility and Global Climate Change. *Leiden Journal of International Law*, **18**, 747–75.

Caney, S. (2006). Environmental Degradation, Reparations, and the Moral Significance of History. *Journal of Social Philosophy*, **37**(3), 464–82.

Caney, S. (2009). Justice and the Distribution of Greenhouse Gas Emissions. *Journal of Global Ethics*, **5**(2), 125–46.

Caney, S. (2010a). Climate Change and the Duties of the Advantaged. *Critical Review of International Social and Political Philosophy*, **13**(1), 203–28.

Caney, S. (2010b). *Justice, Equality and Greenhouse Gas Emissions?* In Unpublished manuscript presented at the conference 'Responsibility in International Political Philosophy' at the University of Graz, Graz, 20 September 2010. [This is an extended version of the article published in 2012 in Philosophy & Public Affairs.]

Caney, S. (2012). Just Emissions. *Philosophy & Public Affairs*, **40**(4), 255–300.

Cripps, E. (2011). *Acknowledging the Elephant: Population, Justice and Urgency.* In Unpublished manuscript presented at the conference "Time Dimensions in the Climate Justice Debate" at the University of Graz, 13–15 September 2010.

Duus-Otterström, G., and Jagers, S. C. (2012). Identifying Burdens of Coping with Climate Change: A Typology of the Duties of Climate Justice. *Global Environmental Change*, **22**, 746–53.

Eyckmans, J., and Schokkaert, E. (2004). An "Ideal" Normative Theory for Greenhouse Negotiations? *Ethical Perspectives*, **11**(1), 5–19.

Gardiner, S. M. (2011). *A Perfect Moral Storm: The Ethical Tragedy of Climate Change*. New York: Oxford University Press.

Gosseries, A. (2007). Cosmopolitan Luck Egalitarianism and the Greenhouse Effect. *Canadian Journal of Philosophy*, **31**, 279–309.

Hartzell-Nichols, L. (2011). Responsibility for Meeting the Costs of Adaptation. *Wiley Interdisciplinary Reviews: Climate Change*, **2**(5), 687–700.

Hayward, T. (2009). International Political Theory and the Global Environment: Some Critical Questions for Liberal Cosmopolitans. *Journal of Social Philosophy*, 40(2), 276–95. DOI: 10.1111/j.1467–9833.2009.01451.x

IPCC. (2007). *Climate Change 2007: Impacts, Adaptation and Vulnerability*. Contribution of Working Group II to the Fourth Assessment Report of the Intergovernmental Panel on Climate Change, ed. M. L. Parry, O. F. Canziani, J. P. Palutikof, P. J. van der Linden, and C. E. Hanson. Cambridge: University Press.

Kumar, R. (2003). Who Can Be Wronged? *Philosophy & Public Affairs*, **31**(2), 99–118.

Locke, J. (1993). *Two Treatise of Government*. In *Cambridge Texts in the History of Political Thought*, ed. P. Laslett. Cambridge: Cambridge University Press.

Meyer, A. (2000). *Contraction and Convergence: The Global Solution to Climate Change*. Schumacher briefing no. 5. Totnes Devon: Green Books for the Schumacher Society.

Meyer, L. H. (2004). Compensating Wrongless Historical Emissions of Greenhouse Gases. *Ethical Perspectives*, **11**(1), 20–35.

Meyer, L. H., and Roser, D. (2006). Distributive Justice and Climate Change: the Allocation of Emission Rights. *Analyse & Kritik. Zeitschrift für Sozialtheorie*, **28**(2), 223–49.

Miller, D. (2008). *Global Justice and Climate Change: How Should Responsibilities Be Distributed?* URL: http://www.tannerlectures.utah.edu/lectures/documents/Miller_08.pdf. Accessed March 3, 2011.

Moellendorf, D. (2011). Common Atmospheric Ownership and Equal Emissions Entitlements. In: *The Ethics of Global Climate Change*. ed. D. G. Arnold. Cambridge and New York: Cambridge University Press, pp. 104–23.

Moellendorf, D. (2014). *The Moral Challenge of Dangerous Climate Change: Values, Poverty, and Policy*. New York: Cambridge University Press.

Nussbaum, M. C. (2007). *Frontiers of Justice: Disability, Nationality, Species Membership: The Tanner Lectures on Human Values*. Cambridge, MA: Belknap Press of Harvard University Press.

Ott, K., and Baatz, C. (2012). Domains of Climate Ethics. In: *Human Health and Ecological Integrity: Ethics, Law and Human Rights*, ed. L. Westra, C. L. Soskolne, and W. Spady. New York: Routledge, pp. 188–200.

Ott, K. et al. (2004). *Reasoning Goals of Climate Change Protection: Specifiation of art. 2 UNFCCC*. Bad Neuenahr-Ahrweiler: Deutsches Umweltbundesamt (UBA).

Page, E. A. (2008). Distributing the Burdens of Climate Change. *Environmental Politics*, **17**(4), 556–75.

Page, E. A. (2012). Give It Up for Climate Change: A Defence of the Beneficiary Pays Principle. *International Theory*, **4**(2), 300–30.

Parfit, D. (1987). *Reasons and Persons*. Oxford: Clarendon.

Pelling, M. (2011). *Adaptation to Climate Change: From Resilience to Transformation*. London and New York: Routledge.

Rickels, W., Klepper, G., Dovern, J., Betz, G., Brachatzek, N., Cacean, S., Güssow, K., Heintzenberg, J., Hiller, S., Hoose, C., Leisner, T., Oschlies, A., Platt, U., Proelß, A.,

Renn, O., Schäfer, S., and Zürn M. (2011). *Large-Scale Intentional Interventions into the Climate System? Assessing the Climate Engineering Debate*. Scoping report conducted on behalf of the German Federal Ministry of Education and Research (BMBF), Kiel Earth Institute, Kiel.

Risse, M. (2008). Who Should Shoulder the Burden? Global Climate Change and Common Ownership of the Earth. HKS Faculty Research Working Paper series RWP08-075. URL: https://research.hks.harvard.edu/publications/workingpapers/citation.aspx?PubId=6074&type=FN&PersonId=170.

Schuessler, R. (2011). Climate Justice: A Question of Historic Responsibility? *Journal of Global Ethics* 7: 261–78.

Schuessler, R. (2014). Equal Per Capita Emissions Defended: A Second-Order Approach. Discussion Paper. URL: https://www.academia.edu/16496860/Equal_Per_Capita_Emissions_Defended_A_Second-Order_Approach.

Seidel, C. (2012). Klimawandel, globale Gerechtigkeit und die Ethik globaler öffentlicher Güter: Einige grundlegende begriffliche Fragen. In *Globale öffentliche Güter in interdisziplinären Perspektiven*, ed. M. Maring. Karlsruhe: Schriftenreihe des Zentrums für Technik- und Wirtschaftsethik am Karlsruher Institut für Technologie 5. Karlsruhe: KIT Scientific, pp. 179–95.

Seidel, C. (2013). Complex Emission Egalitarianism and the Argument from Global Commons. Unpublished manuscript.

Seidel, C. (2014). On 'Imperfect' Imperfect Duties and the Epistemic Demands of Integrationist Approaches to Justice. *Ethics, Policy & Environment*, **17**(1), 39–42.

Sen, A. (2001). *Development as Freedom*. New York: Knopf.

Singer, P. (2002). One World: The Ethics of Globalization. *The Terry lectures*. New Haven, CT: Yale University Press.

Shue, H. (2014). *Climate Justice: Vulnerability and Protection*. Oxford: University Press.

Tickell, O. (2008). *Kyoto2: How to Manage the Global Greenhouse*. London: Zed Books.

United Nations (UN) (1992). *United Nations Framework Convention on Climate Change*. New York.

United Nations Development Programme (UNDP) (2011). *Human Development Report 2011. Sustainability and equity: a better future for all*. New York and Basingstoke: United Nations and Palgrave Macmillan.

Vanderheiden, S. (2004). Justice in the Greenhouse: Climate Change and the Idea of Fairness. *Social Philosophy Today*, **19**, 89–103.

Vanderheiden, S. (2008). *Atmospheric Justice: A Political Theory of Climate Change*. Oxford: University Press.

Vanderheiden, S. (2011). Globalizing Responsibility for Climate Change. *Ethics & International Affairs*, **25**(1), 65–84.

Voget-Kleschin, L. (2013). *Sustainable Food Consumption? Claims for Sustainable Lifestyles in between Normative and Eudaimonistic Issues. The Example of Food Production and Consumption*. Dissertation thesis. University of Greifswald, Germany.

Voget-Kleschin, L. (2013). Employing the Capability Approach in Conceptualizing Sustainable Development. *Journal of Human Development and Capabilities*, **14**(4), 1–20.

Voget-Kleschin, L. (2015). Reasoning Claims for More Sustainable Food Consumption: A Capabilities Perspective. *Journal of Agricultural and Environmental Ethics*, **28**(3), 455–77.

Walzer, M. (1983). *Spheres of Justice: A Defense of Pluralism and Equality*. Oxford: Robertson.

WBGU (2009). Solving the Climate Dilemma: The Budget Approach. Special Report. URL: www.wbgu.de/fileadmin/templates/dateien/veroeffentlichungen/sondergutachten/sn2009/wbgu_sn2009_en.pdf.

9

In the Name of Political Possibility

A New Proposal for Thinking About the Role and Relevance of Historical Greenhouse Gas Emissions

SARAH KENEHAN

Introduction

Global climate change (GCC) is a complex political and moral phenomenon, as concerns of international justice, intra-national justice, and intergenerational justice all need to be identified and addressed. Central to these concerns is determining to what extent a nation's historical contribution to GCC should inform its obligations going forward; this is the question that I will focus on in this chapter. The case that I will make is simple. Our planet is warming at a rapid rate, and we are close to reaching a threshold, which, once crossed, will commit us to a dangerous amount of warming (IPCC Working Group I, 2013: 18). But there is an upshot: at this moment some of that danger can still be mitigated – if we act quickly, and if we act decisively. The argument that I have to offer is developed in this context, and so I assume that acting to prevent this dangerous warming is the most important goal, the sort of goal that we might be willing to sacrifice other goods for in order to secure. As such, my argument is not necessarily one about the ideals of distributive and reparative justice, but rather one that takes seriously and is informed by the very real political hurdles and time constraints that face decision makers and world leaders. To be sure, the conclusions that I reach in this chapter stray from the moral ideal, but they are aimed at moving us a little closer to defining a compromise that is both politically viable and morally defensible.

I begin by detailing some of the arguments and issues that arise on the policy front when policy makers negotiate over the importance of historical contributions to GCC, and I discuss a popular philosophical conception of the moral ideal in this context. There is a notable gap between what political philosophers talk about and what is presently possible to bring about in climate negotiations. Given the urgency of the problem, I propose that what is needed is a morally defensible solution that is informed by the virtue of political feasibility. In section II, I explain this virtue, and I compare the feasibility of three different ways of incorporating nations' historical contributions to GCC into a climate treaty. This analysis shows first that mandating full accountability for historical emissions is very close to being infeasible; second, that offering zero accountability for historical emissions is likely the most feasible option (though it risks cynical realism); and, third, that demanding partial responsibility for historical emissions is at least minimally feasible. In the remaining sections, I argue for an iteration of the latter sort of option (though certainly not the only conceivable iteration) that focuses on the special moral character of post-1990 emissions. I show that, while not morally perfect, my partial responsibility proposal does respect and reflect some important moral values in ways that the zero accountability option fails to do, and it is more feasible than the full responsibility option. As such, I believe that the partial responsibility option is the one that policy makers and political philosophers should pursue going forward, as it strikes an important balance between political feasibility and moral desirability.

I Accounting for Historical Responsibility: Tension between the Ideal and the Real

World leaders have yet to arrive at a consensus regarding the extent to which historical contributions to climate change will inform future obligations. The polluter pays principle ("you broke it; you fix it") is the primary instrument by which policy makers have tried to factor in historical contributions to GCC; it has been both strongly defended and ardently opposed in climate negotiations. In arguing that industrialized nations should carry the burden of the costs of climate change mitigation and adaptation, authors of the "Brazil Proposal" cite numerous times that "the largest share of historical and current global emissions of greenhouse gas has originated in the developed countries" (UNFCCC, 1997: 5, 11, 18). And more recently, the official report from a workshop on equity and sustainable development in the context of climate change noted:

> For some, historical responsibility was central to the discussions on a formulaic approach, and this issue has to be resolved in terms of responsibility for current impacts before discussing the responsibility of non–Annex I Parties. For others, it was not seen as an adequate measure of equity because it is complex, static and includes a large number of variables.[1] *(UNFCCC, 2012: 11–12)*

[1] Similarly: "Regarding historical responsibility, a representative of the United States noted that this is a static concept, and suggested the need for a dynamic approach, based on lessons learned from the past 20 years. Responding to this, a representative of Saudi Arabia noted that Parties should solve the issue of the impact of emissions from the time of the industrial revolution until 1990" (UFCCC, 2012: 10).

So, on the policy level, the extent to which historical contributions should be factored in is surely a matter of contention, and has been for some time now.

To be sure, there are many morally defensible philosophical positions that have arrived at well-reasoned and substantive ways of incorporating historical contributions into present responsibilities.[2] I will assume for the purposes of this chapter that the moral ideal in this context demands full accountability for a nation's historical contribution to GCC, justified most basically by commonsense appeals to fairness.[3] As Singer argues: "To put it in terms a child could understand, as far as the atmosphere is concerned, the developed nations broke it. If we believe that people should contribute to fixing something in proportion to their responsibility for breaking it, then the developed nations owe it to the rest of the world to fix the problem with the atmosphere" (Singer, 2002: 33–34).[4] But while theories that defend this ideal may be reasonable articulations of what justice demands in this context, we may have grave doubts as to whether securing this ideal is possible given the current global political environment.[5] This is not to say that the aforementioned analyses are incorrect or are lacking in all practical relevance; but rather, given the very extreme time constraints that policy makers must work under, combined with the severe and dangerous consequences that will result if we do not achieve climate stability, there is an urgent need for a morally defensible account of historical responsibility that takes seriously the limits of political possibility.[6]

Herein lies the problem: we must strike a balance between securing the moral ideal and proposing and supporting solutions that are politically workable so that we actually can achieve climate stability.[7] It is my view that the latter goal – achieving climate stability – takes precedence over the goal of reaching an agreement that fairly distributes the responsibilities for contending with a warming world. This prioritization is justified by the fact that, left unchecked, the consequences of climate change will be severely damaging, including the devastation of populations, landscapes, and ways of life. That said, this ordering does not necessitate that the second goal is unimportant or superfluous: to the

[2] See, for instance: Caney, 2006; Gosseries, 2004; Meyer, L. H. and Roser, D., 2010; Miller, 2008a; Shue, 1999; Singer, 2002.

[3] Various thinkers have defended this ideal: Gosseries, 2004; Meyer, L. H. and Roser, D., 2010; Shue, 1999; Singer, 2002.

[4] Likewise, "Those societies whose activities have damaged the atmosphere ought . . . to bear sufficiently unequal burdens henceforth to correct the inequality that they have imposed. In this case, everyone is bearing the costs . . . but the benefits have been overwhelming skewed towards that who have become rich in the process" (Shue, 1999: 105).

[5] This claim will be elaborated on in section III when I discuss the feasibility of the Full Responsibility proposal.

[6] Traxler seems to be close to doing this in his argument, though he offers no account of what feasibility demands (see Traxler, 2002).

[7] Of course, this presupposes a particular understanding of moral ideals that is popular, but not universally accepted. For a different understanding, see, for instance, Farelly, 2007.

contrary, reaching a just agreement to achieve climate stability would move us closer to a world in which the relationships among nations more closely reflected the ideals of fairness, equality, and mutual respect, and so securing the second goal would mark important progress toward a more just and connected future. But in the context of the danger and urgency that characterize the problem, we cannot hold out for perfect justice, and we should be willing to think about the acceptability of agreements that move us closer to, but ultimately fall short of, the ideal.

Toward this end, in section II, I will outline the virtue of political feasibility. Then, in the remainder of the discussion, I will articulate and defend a conception of historical responsibility that takes this virtue seriously.

II Political Feasibility

A state of affairs is feasible "if it is one that we could actually bring about" (Gilabert and Lawford-Smith, 2012: 809), that is, "iff there exists an agent with an action in her (its) option set within the relevant temporal period that has a positive probability of bring it about" (Lawford-Smith, 2013: 250). As such, political feasibility involves the relation of agents and aims in particular historical contexts (Gilabert and Lawford-Smith, 2012: 809). Feasibility assessments can be helpful in the evaluation of proposals in at least two ways: they can help to rule out proposals completely (binary determinations), and they can guide comparative assessments of alternative proposals (scalar determinations) (ibid.: 815).

Concerns about political feasibility were first articulated by Rawls, as he sought to develop a theory that not only moved us closer to reflecting the political ideals of fairness, reciprocity, and mutual respect, but also took into consideration what is possible, given the world as it is and the relationships that exist; he called such a theory a realistic utopia. In Rawls' words, a theory that is realistically utopian accepts "men as they are" and "laws as they might be" (Rawls, 1999: 7). Indeed, Rawls thought that the goal of the political philosopher should be to articulate theories and principles that take these constraints seriously (ibid.; see also ibid.: 127–28). Similarly, Brennan and Pettit note that a failure of political philosophers to consider issues of feasibility "represents a potentially serious limitation on the relevance of political philosophy for real-world policy. It suggests that philosophy ought to seek something beyond the purely ideal sort of theory that is fashionable in many circles" (Brennan and Pettit, 2007: 258). They further argue that ignoring the importance of feasibility will render philosophers nothing more than visionaries, focusing on ideal systems with ends that may be impossible or even counterproductive to achieve (ibid., 261) And likewise, Miller contends that while the ideals in political philosophy need not be constrained by an appreciation of empirical facts about human nature and the like, in considering the application of the ideals, we must necessarily consider such limits:

In order to apply these principles and come up with some practical rules for ordering society, we have to bring in factual evidence about the kind of society in which the principles are going to be applied. *(Miller, 2008b: 30)*

Simply put, this means that political philosophers should take political feasibility seriously, and so should seek to identify goals and policies that are actually (if even ambitiously) achievable, not just logically possible.[8]

Accessibility, that is, questions of trajectory and concerns about routes to desired ends, is crucial to the idea of feasibility. Constraints to accessibility can be divided into two types: hard constraints and soft constraints (Gilabert and Lawford-Smith, 2012: 813). Hard constraints include the logical, physical, and biological constraints on a theory; violating these constraints rules out the feasibility of a proposal in the binary sense (ibid.: 813). Economic, institutional, cultural, psychological, and motivational constraints can be classified as soft constraints; a violation of the soft constraints does not necessitate that a proposal is infeasible, but does make the proposal comparatively less so (ibid.). The latter sorts of constraints should not be understood as permanent, and so what is infeasible now may not be so later; this means that we can arrive at different feasibility assessments depending on the time frame we are considering the proposal within (ibid.: 814–15).[9]

Gilabert and Lawford-Smith propose two feasibility tests. The one most relevant to the project at hand, the scalar test, states:

It is more feasible for X to bring about O_1 than for Y to bring about O_2 when it is more probable, given soft constraints, for X to bring about O_1 given that he or she tries than it is for Y to bring about O_2 given that he or she tries. *(ibid.: 815)*[10]

Applying this test is no easy task, and there is no possible way to examine all the possible proposals for considering the relevance of a nation's historical

[8] This is not to say that ideal theorizing is irrelevant or unimportant. Rather, the ideal will be the standard according to which the proposals should be measured (Gilabert, 2012: 45). That is, it will force us to consider whether the compromises embody the right sorts of virtues and values, qualitatively and quantitatively speaking. Moreover, the ideal will give us a starting point from which to critique the status quo, and as such will offer us direction toward "justice enhancement" and "injustice reduction" (ibid.: 47). Similarly, David Estlund argues that feasibility requirements should not shape our ideal theorizing (Estlund, 2011).

[9] Keeping the dynamic nature of soft constraints in mind is crucial for avoiding cynical realism (Gilabert and Lawford-Smith, 2012: 813).

[10] The first test is the binary test; see: Gilabert and Lawford-Smith, 2012: 815. I will assume for the sake of this project that alternative proposals for dealing with historical GHG emissions do not violate any so-called hard constraints and so are not necessarily infeasible.

contribution to GCC in the space permitted.[11] However, we can compare the relative feasibilities of three general types of proposals:

Full Responsibility (FR) – Any global treaty aimed at combating global climate change should include the mandate that a nation is fully responsible for its historical GHG emissions.

Partial Responsibility (PR) – Any global treaty aimed at combating global climate change should include the mandate that a nation is only partially responsible for its historical GHG emissions.

Zero Responsibility (ZR) – Any global treaty aimed at combating global climate change should include the mandate that a nation is not responsible for any of its historical GHG emissions.

In assessing the feasibility of these proposals it is important to identify the agent from whose perspective feasibility is being considered.[12] For the sake of this project, I will assume that the agent in question is the United Nations Framework Convention on Climate Change (UNFCC) Conference of the Parties (COP), that is, the

> supreme decision-making body of the Convention. All States that are Parties to the Convention are represented at the COP, at which they review the implementation of the Convention and any other legal instruments that the COP adopts and take decisions necessary to promote the effective implementation of the Convention, including institutional and administrative arrangements.[13]

Similarly, it is also important to identify a timeline in which we are considering feasibility. The timeline I will assume for this project is 2020, the year that the next UNFCC universal climate agreement is set to be implemented.[14] So the question can be posed as follows: Which of the proposals for historical responsibility – full, partial, or zero – is most feasible?

[11] Indeed, as Gilabert notes: "Given our epistemic limitations, the complete fulfillment of the goal of showing that a proposal is superior to any conceivable (and even prima facie impartially plausible) alternative might not be theoretically feasible," and, "practical decisions have to be made in real time, [so] assessments of alternatives will have to be put on hold in order to make a choice about what to do on the basis of what the inquiry has already provided" (Gilabert, 2012: 44).

[12] As Lawford-Smith notes: "The fact that we get different answers about what is feasible depending on whose project we are interested in is not contradictory. Usually, we will ask feasibility questions from the perspective of the agent whose project is at issue. Consider the difference between asking Kim Jong-un, and what is feasible for Australia, with respect to Australia accepting immigrants from North Korea. It is highly feasible for Kim Jong-un to lift restrictions upon exit, but it is not very feasible for Australia to accommodate North Korean immigrants, because that would require Kim Jong-un to lift restrictions upon North Koreans' exit, and that will not happen anytime soon" (Lawford-Smith, 2013: 256).

[13] United Nations Framework Convention on Climate Change, "Conference of the Parties," <http://unfccc.int/bodies/items/6241.php> (10 June 2014).

[14] The details of this agreement were finalized at the end of 2015. I extended the timeline to 2020 under the assumption that amendments or revisions to the initial agreement are possible. Of course, shortening the timeline will likely affect the feasibility assessments arrived at here.

To answer this question, we have to assess the hard and soft constraints and their effects on each proposal. The hard constraints, that is, the logical, physical, and biological constraints on the proposals, apply equally to each mandate, and none of the proposals appears to violate any of these constraints. However, there are soft constraints that apply differentially to each proposal, thus affecting their relative feasibilities. I will consider the relevant soft constraints for each proposal in the following.

We will consider Full Responsibility (FR) first. Historically speaking, many nations (primarily the wealthy, developed nations) have rejected this mandate, while many other nations (primarily the poor, less developed nations) have pushed for an agreement that embodies this mandate.[15] Of those that would reject this proposal, the United States is probably one of the most harmful historical emitters, and a fierce resistor of such demands.[16] Given the cultural, economic, and political constraints that currently exist within the United States, it is highly unlikely that any treaty that proposed such a measure would be ratified and followed. Support for this claim is found in the fact that a recent study showed that while a majority of Americans (67 percent) believe that global warming is happening, only a little more than half of all Americans believe that, if global warming is happening, it is caused by human activities (Leiserowitz, A. et al. 2015: 3). But, perhaps more importantly, as of June 2013, more than 56 percent (133 members) of Republicans in the House of Representatives and 65 percent (30 members) of the Senate Republican caucus deny that global climate change is occurring.[17] Moreover, historically speaking, the United States has been unable or unwilling to adopt any substantive and sweeping policies to deal effectively with the problem.[18] All of these trends indicate that the United States is very unlikely to accept any treaty that contains the FR mandate. And if the United States is unwilling to accept this sort of treaty, then it is highly unfeasible that an agreement of this type will be adopted by the COP, since, given its size and contribution to the problem, it is imperative that the United States and nations like it are signatories of the treaty, as they have

[15] See section I.

[16] See section I. Consider G. W. Bush's famous remarks "I'll tell you one thing I'm not going to do is I'm not going to let the United States carry the burden for cleaning up the world's air, like the Kyoto Treaty would have done. China and India were exempted from that treaty. I think we need to be more even-handed" (Commission on Presidential Debates, "October 11, 2000 Debate Transcript: The Second Gore–Bush Presidential Debate, URL: www.debates.org/index.php?page=october-11–2000-debate-transcript (10 June 2014)).

[17] Tiffany Germain, "The Anti-Science Climate Denier Caucus: 113th Edition," *Climate Progress*, 10 April 2014, <http://thinkprogress.org/climate/2013/06/26/2202141/anti-science-climate-denier-caucus-113th-congress-edition/> (11 June 2014). According to the study, this means that these representatives "have made public statements indicating that they question or reject that climate change is real, is happening, and is caused by human consumption of fossil fuels" (Germain, "Anti-Science Climate Denier Caucus").

[18] This claim will be discussed in detail in section III.

the potential to cancel out any good that can be done through the reasonable mitigation efforts of the rest of the world.[19]

Of course, nations such as the United States are much more likely to accept an agreement that does not demand responsibility for historical emissions (Zero Responsibility, ZR). Developing nations, on the other hand, would resist a ZR proposal on the basis of serious concerns of fairness: no accountability for historical emissions means that currently developed nations were allowed to emit without restraint and without fear of having to absorb future costs. The dirty, fossil-based path to industrialization taken by the world's currently wealthiest nations allowed them to develop quickly and to establish themselves as economic powerhouses. To demand that they do not have to pay for the costs of their actions, while absorbing most of the benefits, is to place developing nations in a position of subjugation, thus violating the political ideals of fairness, equality, and mutual respect (Kenehan, 2014: 254). Given these concerns, developing nations would likely oppose a proposal that ignored all historical responsibility. But, at the same time, we must also consider the bargaining power that these nations have, relative to the larger, developed nations. Obviously, the ability of most developing nations to shape policy is much weaker than that of the developed nations, and so it is conceivable that the developing nations could be coerced or manipulated into an agreement with which they did not agree.[20] In this way, then, Zero Responsibility seems to be more feasible than Full Responsibility.

Finally, let us consider Partial Responsibility, which mandates that a nation is only partially responsible for its historical contribution to GCC. PR compromises on the issues constraining both the developing nations with regard to Zero Responsibility and the developed nations with regard to Full Responsibility: it lessens the cultural, psychological, and economic burdens of the developed nations and moves toward establishing the equality and mutual respect of the developing nations.[21] But, as it is a middle ground, it does not fully address the shortcomings of the other two proposals. For instance, the cultural and political limitations that exist in the United States with regard to adopting a climate change agreement would still be pertinent here, and the less-developed nations of the world may still feel as though their interests (in, e.g., development and in being politically equal) are not being adequately respected via this compromise. And finally, as before, relative

[19] Consider, for instance, that, in 2011, the United States emitted 19.69 tCO_2 per person, and accounted for 13.4 percent of total global GHG emissions. (World Research Institute, *CAIT2.0 Climate Data Explorer*, <http://cait2.wri.org> (20 June 2014)).

[20] I use the word "most" here since a nation like China would probably carry much weight in shaping policy. In addition, it is also possible that developing nations might group together, and use their collective power to try to influence change. While I agree that would certainly strengthen the position of developing nations as a whole, they would still likely be subject to coercion and manipulation in policy negotiations.

[21] The burdens facing the United States will be described in further detail in section III.

bargaining power needs to be considered: developing nations could be forced into accepting this agreement, despite their objections, or wealthy nations could reject such an agreement (insisting on something closer to ZR), even if developing nations supported it. So far, then, Partial Responsibility seems more feasible than Full Responsibility, but it is unclear how it compares to Zero Responsibility.[22]

A pessimistic comparison of PR and ZR would suggest that ZR is the more feasible of the two, given the sheer power of the nations whose interests would be served by its adoption. But, in assessing feasibility, we should be careful not to slip into a position of cynical realism: "Cynical realism occurs when we surrender to undesirable circumstances we could change in the long-term" (Gilabert, 2012: 50). To prevent cynical realism, Gilabert suggests that we address the softer constraints with particular attention to the possibility of change (ibid.). That is, we should not just accept these constraints as unmovable; rather we should think in the long term while focusing on transitional changes in the short term.[23] To say that ZR is the only chance of adopting a global climate treaty would likely be a case of cynical realism. Aiming for a proposal akin to PR is surely more ambitious, but it is possible that the cultures of major players such as the United States might change and become more accommodating to a fairer climate change agreement with the right sorts of pressure and transitional polices. As such, PR is at least minimally feasible and, importantly, appears to avoid sinking into cynical realism.

In the remainder of the chapter, I will try to show that some versions of the PR proposal also have the potential to embody the characteristic of moral desirability in ways that ZR does not. This virtue, combined with feasibility, offers strong reasons to favor an approach to climate negotiations that adopts a PR type of framework, as it strikes a balance between striving for the moral ideal and taking political feasibility seriously. And so, all things considered, I will argue that this is the optimal sort of approach that policy makers should pursue.

III Historical Responsibility: A Proposal

As noted, there are likely several feasible and morally desirable versions of PR. In this section, I will argue for a specific iteration of this proposal, one that demands historical responsibility from 1990 on, as post-1990 emissions have a special moral character.

[22] Of course, we can imagine that the degrees of partial responsibility asked for can influence the feasibility assessment. It is likely the case that the closer PR is to FR, the less feasible it is (though still more feasible than FR); and the closer PR is to ZR, the more feasible it becomes (though still less feasible than ZR).

[23] Gilabert writes: "We adopt a transitional standpoint focused on dynamic duties, envisioning trajectories of political transformation that approximate the long-term fulfillment of our ambitious principles" (Gilabert, 2012: 50).

1990 Onward: A Case for Moral Responsibility

Moral responsibility can be attributed when an agent acted in "a way that displays moral fault," including deliberate or reckless behavior (Miller, 2007: 100). For the case at hand, moral responsibility for a nation's historical contribution to GCC can reasonably be attributed from around 1990, when the first IPCC report was published. In this report, scientists concluded that "increases in atmospheric concentrations of greenhouse gases may lead to irreversible change in the climate" (IPCC Working Group II, 1990: 1). And:

We are certain of the following:

- there is a natural greenhouse effect which already keeps the Earth warmer than it would otherwise be;
- emissions resulting from human activities are substantially increasing the atmospheric concentrations of the greenhouse gases carbon dioxide, methane, chlorofluorocarbons (CFCs) and nitrous oxide. These increases will enhance the greenhouse effect, resulting on average in an additional warming of the Earth's surface. The main greenhouse gas, water vapor, will increase in response to global warming and further enhance it. (IPCC Working Group I, 1990: xi)

So, as early as 1990, scientists were converging on the conclusion that human activities were dramatically altering the earth's climate. And, just as important, this information was made public and accessible to policy makers and lay people all over the world. This means that from 1990 on, nations continuing down a path of excessive emissions knew of the harm to which they were contributing. Further assessment reports confirmed these conclusions; as such, claims of ignorance from this point on are difficult to ground. Thus, the wealthy, developed nations of the world are morally responsible for their post-1990 GHG emissions.

Moreover, during this time (1990 on), it was not just that most developed nations did nothing to mitigate the harm they were causing (by, e.g., trying to hold their emissions stable or even lessen them); they actually intensified the release of GHG emissions. Particularly, as noted previously, GHG emissions have increased 24 percent since 1990, including a 28 percent increase in CO_2 and an 11 percent increase in CH_4 emissions (IPCC Working Group III, 2007: 101–02). And:

> The annual CO_2 concentration growth rate was larger during the last 10 years . . . than it has been since the beginning of continuous direct atmospheric measurements....The predominant sources of the increase in GHGs are from the combustion of fossil fuels. Atmospheric CO_2 concentrations have increased by almost 100 ppm in comparison to its preindustrial levels, reaching 379 ppm in 2005, with mean annual growth rates in the 2000–2005 period that were higher than those in the 1990s. *(IPCC Working Groups I, II, and III, 2007: 37)*

Total GHG emissions – 1990–2012	Total CO2 emissions – 1990–2012
Australia – increased emissions by 2.4%	Australia – increased emissions by 1.8%
Canada – increased emissions by 42.2%	Canada – increased emissions by 50.5%
United States – increased emissions by 2.7%	United States – increased emissions by 3.0%[1]

Figure 9.1 Australia, Canada, and United States: Increase of Emissions since 1990.

Of course, some of this increase might be the result of the industrialization of developing nations and so may be justified in so far as these emissions are needed to secure goods such as food, power, and the like.[24] However:

For Annex I non-EIT [economies in transition] Parties, from 1990 to 2011 GHG emissions increased by 3.2 percent excluding LULUCF and by 2.1 percent including LULUCF [land use, land use change, and forestry]. *(UNFCCC, 2013: 8)*

And, the three largest historical culprits actually increased emissions since 1990 (Figure 9.1):

The trends exhibited in these numbers are very important. They show that not only did the wealthy, developed nations fail to do something about the problem, but they actually made the problem worse, in full knowledge of the harm that they were contributing to, thus strengthening the claims of moral responsibility.[25]

Simply put, the wealthiest nations of the world that have contributed the most to GCC by way of their historical GHG emissions knew of the harm that they were causing as far back as 1990. This gives good reason to hold them morally accountable for their post-1990 emissions, as the moral character of these emissions differs greatly from that of pre-1990 emissions (a period defined by relative ignorance of the harm to which these nations were contributing).[26] As such, the 1990 threshold provides a non-arbitrary and morally relevant benchmark for thinking about a nation's historical contribution to GCC in the recent past.

Can Historical Trends Be Justified?

From a moral point of view, we might be tempted to dismiss the claims of moral responsibility leveled here if the excessive patterns of emissions exhibited by the world's wealthiest nations can be justified. In addressing this concern, it is important to note there is a huge disparity – historically and currently – between

[24] On this point, consider Shue's argument that we should reject the homogenization of GHGs, as there is a moral difference between emissions being used to secure necessities and emissions being used to produce luxuries and the like (Shue, 1993).

[25] See, for instance Shue, 2010. Of course, this trend does not help to ground the cutoff point that I have proposed; it only shows that wealthy, developed nations were increasingly negligent over time.

[26] Of course, a strong case can likely be made for attributing causal responsibility for pre-1990 emissions.

the volume of GHGs emitted by the wealthiest nations and the volume of GHGs emitted by the least well-off nations:[27]

Developed countries (UNFCCC Annex I countries) hold a 20% share in the world population but account for 46.4% of global GHG emissions. In contrast, the 80% of the world population living in developing countries (non–Annex I countries) account for 53.6% of GHG emissions.[28] *(IPCC Working Group III, 2007: 106)*

Unfortunately, space here does not permit a complete investigation of how emissions between 1990 and the present have been used, but we can look at some important turning points over the two-plus decades for one of the largest culprits of excessive emissions – the United States.[29]

In 1992, the United States, under the leadership of President G. H. W. Bush, ratified the UNFCCC, while refusing to commit the country to binding emissions reductions. In lieu of this commitment, President Bush submitted the EPAct (Energy Policy Act) of 1992, which, while having some beneficial effects with regard to GHG reductions, was primarily aimed at lessening the nation's dependence on foreign energy supplies (Parker, Blodgett, and Yacobucci, 2011: 3). *The National Action Plan for Global Climate Change*, submitted late that same year, listed voluntary measures to reduce U.S. GHG emissions, with the aim of reducing them to near 1990 levels by 2000; actual levels in 2000 were 14.3 percent higher than 1990 levels (ibid.: 5 and 5 n. 12).

A second *Climate Action Plan* was submitted by President Clinton in 1993. While the goal outlined in this plan was ambitious (i.e., restore the nation's GHG emissions to 1990 levels), reductions were, again, voluntary; this goal was abandoned in 1997 upon the realization that the nation's emissions were actually increasing (ibid.: 6). Kyoto negotiations began that same year, and the resulting Kyoto Protocol was signed by President Clinton. However, this victory was short lived, as Congress passed resolutions proactively rejecting the protocol's ratification on the basis that developing nations were not held to the same emissions schedules as the developed ones (ibid.: 7).

Kyoto negotiations were left completely behind upon the election of President G. W. Bush, and even though the nation remained a party to UNFCC, the United States altogether abandoned the goal of achieving 1990 levels of emissions. Instead, in 2002, President G. W. Bush implemented a voluntary program that

[27] China is the current exception to this trend.

[28] Similarly, "Differences in per capita income, per capita emissions and energy intensity among countries remain significant. In 2004, UNFCCC Annex I countries held a 20% share in the world population, produced 57% of the world's Gross Domestic Product based on Purchasing Power Parity (GDP_{PPP}) and accounted for 46% of global GHG emissions" (IPCC Working Groups I, II, and II, 2007: 37).

[29] I am looking at only 1990 on because I assume that most nations would have no reason to change their emissions behaviors substantially previous to the publication of the first IPCC report.

focused on reducing "the intensity of emissions per unit of economic activity" (ibid.: 8). The goal of the plan was to reduce GHG intensity by 18 percent, though this was projected to occur simultaneously with an increase in overall GHG emissions (ibid.) Two EPActs followed – one passed in 2005 and another passed in 2007 – and each lacked direct climate stabilization initiatives (ibid.: 9).

Under the direction of President Obama in 2009, the nation appeared to be gathering momentum toward adopting effective and substantive climate policy leading up to the Copenhagen Conference (ibid.: 10–11). At the end of 2009, the United States signed on to the Copenhagen Agreement, which dictated lofty, but non-binding goals. That same year, the American Recovery and Reinvestment Act was signed into law. As part of the act, the United States invested more than $31 billion toward a variety of clean energy projects.[30] In the spring of 2014, President Obama presented his *Climate Action Plan*, detailing methods to reduce U.S. GHG emissions to 17 percent below 2005 levels by 2020.[31] But, despite this movement forward, the United States has not yet formally agreed to the Doha Amendment (proposed in 2012), which would extend the commitment period of the Kyoto Protocol and set binding post-2020 targets for member parties of the UNFCC.[32]

So, clearly, there were many opportunities for the United States to change its emissions behavior, and yet, the effectiveness of the policies that has enacted has been minimal, at best: as noted, since 1990, the United States has actually increased both its CO_2 emissions and its overall level of GHG emissions.[33] The vast majority of the measures implemented by the nation's leaders were voluntary and/or relatively insignificant, as the foundational concern underlying most of these acts and plans was not reducing the nation's contribution to GCC, but rather minimizing the implementation costs of GHG reduction measures and reducing the effects that that these measures would have on the nation's competitiveness in the global market (Parker, Blodgett, and Yacobucci, 2011: summary). Thus, a case can be made that the United States never made a real and serious effort to limit its contribution to GCC until 2009. And, even then, many of the plans proposed in

[30] United States Department of Energy, "Recovery Act," *Energy.gov* <http://energy.gov/recovery-act> (1 December 2013). This included $602 million for residential credit for alternative energy, $647 million for electricity produced from renewable resources, $142 million for the extension of commuter transit benefits, $144 million in business credits for renewable energy, and $51 million in increased credits for alternative fuel vehicle refueling properties. ("Overview of Funding," "The Recovery Act," *Recovery.gov: Track the Money* <www.recovery.gov/Transparency/fundingoverview/Pages/default.aspx> (1 December 2013)).

[31] United States Department of State, "Fact Sheet: 2014 U.S. Climate Action Report," <www.state.gov/e/oes/rls/rpts/car6/219259.htm> (12 June 2014).

[32] United States Framework Convention on Climate Change, "Status of Ratification of the Protocol," *Kyoto Protocol*, <http://unfccc.int/kyoto_protocol/status_of_ratification/items/2613.php> (3 December 2013); and United Nations Framework Convention on Climate Change, "Doha Amendment," *Kyoto Protocol*, <http://unfccc.int/kyoto_protocol/doha_amendment/items/7362.php> (3 December 2013).

[33] Though, for claims of moral responsibility, it would be sufficient to note just one opportunity to change.

2009 and after will take years to implement; as such, the full effectiveness of these measures has yet to be determined.

In addition, many of the emissions released by the United States since 1990 were not, in fact, necessary for securing basic human needs or even for securing the prevailing standard of living. The changes in infrastructure, economic activity, and personal behavior that are likely needed for the United States to reduce its GHG contribution are all thought to pose a threat to a particular American "way of life." Such a lifestyle is marked by big homes, big cars, and a diet based primarily on the consumption of animal products, goods, and activities that are all carbon intensive.[34] While surely some emissions are necessary to sustain shelter, transportation, and nutritional needs, it is undoubtedly the case that current rates of emissions extend well beyond securing these necessities. As a matter of comparison, consider the fact that Germany – a nation similarly developed with a comparable standard of living – managed to achieve a 23.5 percent reduction in its GHG emissions between 1990 and 2012, simultaneously maintaining its high standard of living, and this trend is likewise true for many other similarly developed countries.[35] As such, it is unlikely that the emissions trends of the United States (and other similarly situated nations) can be justified on the grounds that they are necessary for securing subsistence needs, and the like.

Thus, to the extent that the opportunities of other wealthy, developed nations to change their destructive emissions habits mirrored those of the United States, and to the extent that these excessive emissions behaviors were not needed to secure subsistence needs, we can conclude that the intensity of the post-1990 emissions behaviors of the United States and similarly situated nations cannot be justified. As such, from 1990 on, those nations that continued down a path of excessive emissions bear moral responsibility for their actions, and this attribution of moral responsibility grounds, very strongly, the claim that wealthy nations should pay for the costs of their actions.[36]

In the sections that follow, I will consider the moral desirability of the particular PR proposal that I articulated here, and I will consider this in light of the previous feasibility assessment. I hope to show that while the iteration that I have defended

[34] Consider, for instance, that in 2011 transportation emissions accounted for 25 percent of the nation's total GHG emissions (World Research Institute). And between 1990 and 2012, 31 percent of the nation's methane emissions were produced through animal agriculture (United States Environmental Protection Agency, "Overview of Greenhouse Gases: Methane Emissions," 31 July 2013, <epa.gov/climatechange/ghgemissions/gases/ch4.html> (4 August 2013)).

[35] United Nations Framework Convention on Climate Change, "GHG emission Profiles." Of course, this trend is not the rule. Important exceptions include Canada and Australia; as noted previously, both of these nations, like the United States, have increased their emissions since 1990.

[36] Moral responsibility may also demand more than this. This will be discussed in section IV. Other sorts of responsibility that might support this same claim include outcome responsibility, causal responsibility, and remedial responsibility. See, for instance, Miller, 2007.

is not morally perfect, it does strike an important balance between exhibiting political feasibility and movement toward the moral ideal.

IV Moral Desirability and Securing a More Just Future

From sections I and II, it appears that, in the context of GCC, there is an inverse relationship between achieving the moral ideal and arriving at proposals that are politically feasible: the closer we move toward achieving the ideal, the less feasible the proposal becomes, and the more feasible the proposal, the less moral appeal it has.[37] Thus, what is needed is a compromise that strikes a balance between these two goals. I have already shown that my proposal exhibits political feasibility; in this section I hope to show that the sort of compromise embodied by this iteration preserves some morally important characteristics and moves us closer to securing the ideals of justice and fairness in the future, even if it falls short of moral perfection.

The unstated route that my theorizing on this issue has taken emphasizes forward-looking considerations (considerations that prioritize movement toward an ideally just state) over backward-looking considerations (considerations that seek to restore the parties to some status quo). This sort of reasoning fits nicely with the aims previously stated – the achievement of climate stability and the just distribution of the burdens of a warming world – but it also contextualizes climate negotiations as an opportunity to move nations closer to achieving just and equal relationships. Movement toward this goal counts as a moral good, and so the ability of each of the proposals to do this should be examined.

As argued, FR proposals currently appear to lack feasibility. And, although this is the moral ideal, if nations continue to insist on treaties that demand full responsibility in the current political climate, that course will only serve to reinforce the damaging power differentials that exist between nations when the wealthiest nations inevitably refuse to comply, thus creating even greater obstacles to securing both a stable climate and a just future. As such, trying to implement FR in this context would be counterproductive to achieving the moral ideal. Likewise, ZR proposals that ask for a clean historical slate and that give zero weight to past behaviors may be politically workable, but they are, at the same time, a potential threat to a more just future, as they ignore the political virtues of fairness, equality, and reciprocity, and, again, reinforce damaging power differentials in that they appeal primarily to the interests of the developed nations, while minimizing the interests of the undeveloped

[37] I am not claiming that this relationship is necessary, in this context, or in others; other issues in other contexts might reveal a different relationship between exhibiting political feasibility and achieving the moral ideal.

nations. To the extent that a respect for these values is needed in securing a more just future, setting them aside can threaten this possibility. The middle ground – a PR proposal of the sort that I have articulated – takes seriously not only feasibility concerns, but also moral concerns. It holds the damaging emitters morally responsible for at least some of their actions and in doing so takes the demands of the less powerful, developing nations seriously, thus moving us closer to securing the political ideals of fairness, equality, and mutual respect going forward.

This is not to say that my proposal is without moral costs. True causal responsibility for GCC can surely be attributed to most wealthy, developed nations as far back 1800 (earlier or later, depending on the specific nation in question), and so by reaching only as far back as 1990, I am setting aside nearly 190 years of harm causing behavior, which may compromise the aforementioned political values. But the advantage of this compromise is in the form of increasing feasibility. Compare, for instance, the different time frames involved in FR and my version of PR. The iteration of PR that I have defended asks us to take responsibility for actions extending back to 1990, while FR asks us to bear responsibility for actions as far back as the beginning of the Industrial Revolution. Importantly, almost three decades into the past (1990) is not a difficult reach for individuals to make. We can easily place our parents or ourselves at events twenty-seven years prior; we hear stories and see pictures of people we know from twenty-seven years ago, and we have friendships and relationships twenty-seven years old; in short, twenty-seven years into the past is an amount of time that is readily susceptible to human grasp, imagination, and relation. This relatability, gained through sacrificing some moral goods, makes the time frame for grounding responsibility much more tangible and accessible, and so makes the proposal more feasible. To state it more generally, some moral sacrifices are necessary to achieve political feasibility.

V Balancing Political Feasibility and Moral Desirability

In light of the analyses in the preceding sections, we are now in a position to revisit the comparative feasibility between ZR and the specific iteration of PR that I have defended. As noted in section II, claiming that ZR is the only feasible option commits us to cynical realism, which is something that we want to avoid. Nonetheless, ZR is still more politically feasible than PR, considering the short time frame we are working under, and considering all of the soft constraints discussed previously. But given the dual aims of securing feasibility and moving closer to the moral ideal, feasibility is not the only consideration that matters. So, while we have rejected FR as the least feasible option, we should nonetheless try to establish movement toward the ideal. As such, it is important for the proposal dictating historical responsibility to reflect as many morally desirable characteristics as possible, while still maintaining political feasibility.[38] Earlier, I showed that PR is a politically workable solution, and in the previous section I showed that my version of PR reflects the moral values of fairness, equality, and mutual respect, and so will move us closer to securing the ideals of justice in ways that ZR will not. In comparison, politically feasible ZR denies the moral ideals of fairness, equality, and mutual respect, and so compromises the abilities of nations to secure just relationships in the future. As such, the version of PR that I have defended strikes a crucial and necessary balance between feasibility and securing the ideal, in ways that ZR is not able to do.

Importantly, this is not to say that the iteration of PR that I have articulated is the very best version of PR that we can imagine. It is surely possible that other thinkers can defend claims of partial responsibility that reflect the political virtues of fairness, equality, and mutual respect just as well as, or better than, my proposal, and that are, at the same time, politically feasible.[39] Rather, my intention was to show that claims of partial historical responsibility can be politically feasible, and while they likely fall short of moral perfection, they can nonetheless still exhibit some important moral values; in short, it is possible to imagine a morally defensible compromise that is also politically viable.

And, of course, there are versions of PR that are more feasible than others. The specific iteration of PR that I have defended asks historical emitters to be morally responsible for the costs of their actions from 1990 on. Moral responsibility may demand not only undoing the harm that was done, but also providing some sort of reparations, that is, an acknowledgement of moral accountability (an apology, monetary payments above and beyond the cost of the damage done, etc.) This is justified because not only did the emissions contribute to a tangible harm, but the harm causing actions were committed in full knowledge of the damage that they were contributing to, and so there was also injury done to the moral relationship(s) between the victim(s) and the wrongdoer(s).[40] Clearly, if we decide that

[38] As Gilabert explains, the ideal will be the standard according to which the proposals should be measured. Taking the ideal seriously will force us to consider whether the compromises embody the right sorts of virtues and values, qualitatively and quantitatively speaking (Gilabert, 2012: 45 and 47).

[39] For instance, Meyer and Roser defend the following: "Based on the unequal economic progress of countries certain parts of past emissions should be taken into account for the purpose of distributing emissions rights; namely, those past emissions that occurred during the lifetime of the presently living and those past emissions that were side products of benefits which are still around today" (Meyer and Roser, 2010: 240).

[40] We can contrast this to wrongless harms, which are not morally blameworthy. As Reidy and von Platz note: "Reparative claims, on the other hand, are made in the name of victims and for the sake of their moral relationship to wrongdoers... They are predicated on wrongdoing or injustice. They give rise to compensatory demands to the extent that the wrongdoing itself imposed a harm or loss and compensation is necessary or conducive to making amends and repairing moral relations between the parties. But it is the wrong that is fundamental... Even when reparative claims demand compensation, as they often do, they never demand only compensation. An apology or some further reparative act is always required. Reparative justice aims not at a just distribution of the costs of various harms or losses. It aims rather at the repair of moral relationships broken by wrongdoing or injustice. Indeed, as a response to a reparative justice claim, mere compensation can be

moral responsibility in this context dictates monetary payments above and beyond the costs of the harms themselves (or something similar) – as opposed to some non-monetary form of reparations or no reparations at all – then this needs to be factored into the feasibility equation. While respecting this mandate will move us closer to respecting the ideal of justice in this context, it will simultaneously lessen the feasibility of this proposal, at least in the short term.[41] As such, a more feasible version of my proposal would be to excuse demands for reparations, insofar as they ask for monetary payments above and beyond the costs of the actual harms and the like.[42]

Indeed, as we near the dangerous 2°C threshold, the scales tip in favor of those options that exhibit greater feasibility. Again, justification for this sacrifice is grounded in the priority ordering of the goals: achieving climate stability is paramount to achieving the ideals of distributive and reparative justice in this context. This likely means, then, that the closer that we approach surpassing the dangerous threshold, the less feasible PR becomes, and the more feasible ZR becomes; this is especially the case if we have made no progress at chipping away at the current soft constraints, or if they become even more difficult to navigate.

To be clear, in this section, I hope to have shown that the PR proposal that I offered strikes a crucial balance between feasibility and moral desirability, given our current political climate. As feasibility is a function of the relationship of agents and aims in particular historical circumstances, feasibility assessments can change, and so should our assessment of the balance that we seek to strike between feasibility and moral desirability (Gilabert and Lawford-Smith, 2012: 809). So, while the iteration that I have defended is workable and desirable now, in our current global political climate, alternative iterations may become more or less feasible in the future, depending upon the progress (or lack thereof) that policy makers make toward removing or lessening the influence of the relevant soft constraints.

offensive, an assertion that, while the victim might have been harmed, she was not wronged" (Platz and Reidy, 2006: 361–62).

41 This assessment is based on the historical track record of the largest, most damaging emitters detailed previously: if they have thus far expressed an unwillingness to be responsible for the harms they committed, it is very unlikely that they will be willing to pay for the costs of the moral damage that they committed.

42 Of course, this compromise is not without significant costs, as failing to offer proper reparations for a harm can have significant consequences. For instance, it may be that the historical wrong can have such an impact as to constitute an important part of a people's common history and self-identity. Understood this way, to treat the rectification as unimportant or secondary may be disregarding the self-identity and, in some ways, the right to self-determination of a nation. See, for instance, Waldron, 1992. In addition, it may be that allowing such an action to remain unrepaired would be an example of free riding, and so would be an infringement on the political ideals of social utility, fairness, reciprocity, and the like. Reparations would go a long way in repairing these transgressions.

Conclusion

For the purpose of this chapter, I have assumed that there are two goals we should seek to secure via climate negotiations: the first is to achieve climate stability, and the second is to distribute the costs of GCC in a way that secures the moral ideal of justice in this context. I have argued that because of the dangerous and often irreversible consequences of GCC, acting to prevent further warming of the planet is a good of such weighty importance that we should be willing to sacrifice other goods in order to secure it; that is to say, this goal is primary to the second goal. As such, there are two corresponding virtues that any proposal offered to contend with the problem of GCC should take seriously: political feasibility and moral desirability, respectively.

In the previous sections, I have argued for a proposal that dictates only partial responsibility for historical emissions. Specifically, I have defended a version of the partial responsibility mandate that holds historical emitters morally responsible for all post-1990 emissions, as these emissions have a special moral character. I showed that such a proposal is more feasible than one that demands complete responsibility for a nation's historical contribution to climate change, as the full accountability mandate is likely unfeasible (given the time frame in which we have to act and the current global political environment). And while my proposal is less feasible than one that excuses all historical responsibility, resorting to such a solution likely yields to cynical realism; fails to reflect the moral ideals of fairness, equality, and mutual respect; and moves us further from securing a just future, thus neglecting the virtue of moral desirability. Stated succinctly, partial responsibility mandates like the one I have defended strike a crucial balance between achieving political feasibility and striving toward the moral ideal.

In conclusion, if we agree that achieving climate stability is one of the most important political and moral problems that we currently face, then we should be ready to make political and moral compromises to achieve this goal. I have offered one such proposal, though there are likely many others. And while there are many of us (myself included) who believe that historical emitters should, ideally, be motivated to do what is right simply because it is right, we need to acknowledge that, if securing climate stability in the time we have left to act is our primary obligation, then political realities will limit our capacity to pursue the moral ideal of justice in this context. As such, philosophers and politicians should start thinking about the compromises that strike an acceptable balance between political feasibility and moral desirability. I have shown that there are at least some partial responsibility mandates that are capable of securing this balance in a way that the alternative options cannot; so, all things considered, proposals dictating only partial responsibility are the ones that policy makers and political philosophers should currently pursue.

References

Brennan, G., and Pettit, P. (2007). The Feasibility Issue. In *The Oxford Handbook of Contemporary Philosophy*, ed. F. Jackson and M. Smith. Oxford: Oxford University Press, pp. 258–79.

Caney, S. (2006). Environmental Degradation, Reparations, and the Moral Significance of History. *Journal of Social Philosophy*, **37**(3), 464–82.

Estlund, D. (2011). Human Nature and the Limits (if Any) of Political Philosophy. *Philosophy and Public Affairs*, **39**(3), 207–237.

Farelly, C. (2007). Justice in Ideal Theory: A Refutation. *Political Studies*, **55**(4), 844–64.

Germain, T. (2014). The Anti-Science Climate Denier Caucus: 113th Edition. Climate Progress, 10 April. URL: http://thinkprogress.org/climate/2013/06/26/2202141/anti-science-climate-denier-caucus-113th-congress-edition/.

Gilabert, P. (2012). Comparative Assessments of Justice, Political Feasibility, and Ideal Theory. *Ethical Theory and Moral Practice*, **15**(1), 39–56.

Gilabert, P., and Lawford-Smith, H. (2012). Political Feasibility: A Conceptual Exploration. *Political Studies*, **60**, 809–25.

Gosseries, A. (2004). Historical Emissions and Free-Riding. *Ethical Perspectives*, **11**(1), 36–60.

Kenehan, S. (2014). Rawls, Rectification, and Global Climate Change. *Journal of Social Philosophy*, **45**(2), 252–69.

Lawford-Smith, H. (2013). Understanding Political Feasibility. *The Journal of Political Philosophy*, **21**(3), 243–59.

Leiserowitz, A. et al. (2015). *Climate Change in the American Mind: Americans' Global Warming Beliefs and Attitudes, March 2015*. New Haven, CT: Yale Project on Climate Change Communication.

Meyer, L. H., and Roser, D. (2010). Climate Justice and Historical Emissions. *Critical Review of International Social and Political Philosophy*, **13**(1), 229–53.

Miller, D. (2007). *National Responsibility and Global Justice*. Oxford: Oxford University Press, 100.

Miller, D. (2008a). Global Justice and Climate Change: How Should Responsibilities be Distributed? The Tanner Lecture on Human Values. Tsinghua University, Beijing, March 24–25. URL: http://tannerlectures.utah.edu/_documents/a-to-z/m/Miller_08.pdf.

Miller, D. (2008b). Political Philosophy for Earthlings. In *Political Theory: Methods and Approaches*, ed. D. Leopold and M. Stears. Oxford: Oxford University Press, pp. 29–48.

Parker, L., Blodgett, J., and Yacobucci B. D. (2011). *U.S. Global Climate Change Policy: Evolving Views on Cost, Competitiveness, and Comprehensiveness*. Washington, DC: Congressional Research Service.

Rawls, J. (1999). *The Law of Peoples*. Cambridge, MA: Harvard University Press.

Shue, H. (1993). Subsistence and Luxury Emissions. *Law and Policy*, **15**, 39–59.

Shue, H. (1999). Global Environment and International Inequality. *International Affairs*, **75**(3), 531–45.

Shue, H. (2010). Deadly Delays, Saving Opportunities: Creating a More Dangerous World? In *Climate Ethics*, ed. S. M. Gardiner, S. Caney, D. Jamieson and H. Shue. Oxford: Oxford University Press, 146–62.

Singer, P. (2002). *One World: The Ethics of Globalization*. New Haven, CT: Yale University Press.

Traxler, M. (2002). Fair Chore Division for Climate Change. *Social Theory and Practice*, **28**, 101–34.

UNFCCC Ad hoc Group on the Berlin Mandate (May 30, 1997). *Implementation of the Berlin Mandate: Additional Proposals from Parties*. Bonn.

UNFCCC Ad Hoc Working Group on Long-Term Cooperative Action under the Convention (August 15, 2012). *Report on the Workshop on Equitable Access to Sustainable Development*. Doha.

UNFCCC (2013). *National Greenhouse Gas Inventory Data for the Period 1990–2011*. Warsaw.

von Platz, J., and Reidy, D. (2006). The Structural Diversity of Historical Injustices. *The Journal of Social Philosophy*, **37**(3), 360–76.

Waldron, J. (1992). Superseding Historic Injustice. *Ethics*, **103**, 4–28.

Working Group I to the First Assessment Report of the Intergovernmental Report on Climate Change, Climate Change (1990). *The IPCC Scientific Assessment: Policymakers Summary*. Cambridge: Cambridge University Press.

Working Group II to the First Assessment Report of the Intergovernmental Report on Climate Change, Climate Change (1990). *Impacts Assessment of Climate Change*. In *Policymakers Summary*. Cambridge: Cambridge University Press.

Working Groups I, II and III to the Fourth Assessment Report of the Intergovernmental Report on Climate Change, Climate Change (2007). *Synthesis Report*. Cambridge: Cambridge University Press.

Working Group III to the Fourth Assessment Report of the Intergovernmental Panel of Climate Change, Climate Change (2007). *Mitigation of Climate Change*. Cambridge: Cambridge University Press.

Working Group I to the Fifth Assessment Report Intergovernmental Report on Climate Change, Climate Change (2013). *The Physical Science Basis, Summary for Policymakers*. Cambridge: Cambridge University Press.

10

Right to Development and Historical Emissions

A Perspective from the Particularly Vulnerable Countries

MIZAN R. KHAN

Introduction

Climate change is the poster child of global diplomacy today. Since the end of the cold war perhaps no other issue has demanded so much time and energy and so many resources from the global community. Still it remains intractable as ever. The roots of this intractability can be traced both to conceptual and to practical aspects of the issue. Climate change is the result of impersonal activities undertaken by every individual and nation-state of the world. Unlike the ozone regime, which touched a limited number of chemicals for which substitutes were already available from some companies and the impacts were likely to affect the citizens of industrial countries more, the climate regime touches the very foundations of modern life and its comforts. Energy production, based predominantly on fossil fuels and their multifarious uses,are at the core ofthe problem. All the activities are done without intentional harm in mind. Moreover, there are temporal and spatial dimensions. It is a stock, rather than a flow pollution problem. So, the question of historical emissions since the Industrial Revolution stands at the core of distributional equity in burden sharing over reduction of greenhouse gas (GHG) emissions.[1] It means that already deposited GHGs will cause some degree of warming; it is already causing it. Again, yesterday's emissions from industrial countries are being mixed with today's rapidly growing emissions from the developing countries, particularly from their major emitters. Besides, the impacts will manifest themselves fully in the decades to come, and future generations are likely to suffer most. But some groups of countries, such as the least developed and small island developing states (LDCs and SIDS), recognized by the UNFCCC as the particularly vulnerable

[1] Different estimates suggest that around three-fourths of the cumulative global emissions were caused by the industrial countries (Brazil's formula for calculating historical responsibility) (cf. UNFCCC, 1997; La Rovere et al., 2002; Gardiner and Hartzell-Nichols, 2012), while another source mentions that during 1860–2010, the share of U.S. cumulative emissions stood at 19 percent, the European Union's at 17 percent, and China's at 12percent (Stavins,2014).

countries (PVCs), will be and are already being hit first and hardest. Continued undersupply of mitigation already is wreaking havoc on the economies and societies, especially in the PVCs. Scientists attribute the trend of increased magnitude, frequency, and severity of climate disasters of recent years to climate change (IPCC, 2007a, 2012). Then the main agents causing the problem are the rich industrial countries, which are likely to suffer less, while the poor, with the least contribution to the problem, will suffer the most. Some scholars argue that of the most vulnerable coastal regions, the Ganges–Brahmaputra basin stands out (Patz et al., 2005).Though the industrial countries also are suffering from the climate disasters, they have better adaptive capacityand are investing billions in enhancing their adaptive capacity, what Desmond Tutu called an 'adaptation apartheid'. This might be the reason why in the case of climate change the concept of 'vulnerability interdependence' (Volger, 1995:198) is not effective yet. But more focus on adaptation in the industrial countries without ambitious mitigation is likely to make the PVCs worse off (Michaelowa, 2001).

There is no disagreement in the UNFCCC negotiations that mitigation is the ultimate solution, but disagreement over distributional equity in terms of who will do what and how much and how the science-determined emissions allowance within the remaining atmospheric sink capacity will be distributed among the convention parties. The Conference of the Parties Twenty One (COP21) of the UNFCCC held in Paris in early December 2015 reached a universal agreement, as stipulated under the Durban Platform (DP), but it is interesting to note that mitigation responsibilities of parties remain voluntary, while the procedural part, such as periodic submissions of nationally determined contributions (NDCs) and five-year review, will remain mandatory (decision 1/CP.21).The outcome of the Paris conference can be said to vindicate the U.S. position of a legal symmetry that the major emitters from the developing world have to assume such responsibility for binding reductions. But the debate over absolute reduction by the industrial countries versus the reduction of emission intensity by the emerging major emitters including China was not resolved in Paris. Soit remains doubtful that an ambitious mitigation regime with the agreed bottom-up approach even after the Paris Agreement comes into force will be implemented.

Against this backdrop, the right to promote development and sustainable development by parties as an essential condition for addressing climate change has been recognized in the Convention Article 3.4. The Preamble of the convention begins with "Acknowledging that change in the Earth's climate and its adverse effects are a common concern of humankind," but paragraph 3 of the Preamble refers to disproportionate 'per capita' and 'historical emissions' of the industrial countries. Again in the latter part, the Preamble emphasizes integrated understanding of sustainable development that combines the economic, social, and environmental dimensions.

The PVCs are already suffering from both development and adaptation deficits. So their immediate concern is how to strengthen efforts aimed at both these fronts. A sizable literature on justice and human rights argues that climate change violates basic human and development rights (Okereke, 2008; Caney, 2006; Shue, 1992, 1999; Page, 2008). Another strand of literature deals with the ethical principles of distributional equity (Meyer and Roser, 2006; Okereke, 2008; Baer and Sagar, 2010; Baer et al., 2000; Duus Otterstrom and Jagers, 2012; Moellendorf, 2012; Page, 2008; Posner and Weisbach, 2010; Vanderheiden, 2008, 2011). Also there is an array of policy proposals on how to appropriate the remaining atmospheric sink capacity as a global commons (Roberts and Parks, 2007; La Rovere et al., 2002; Clausen and McNeilly, 1998).But no progress has occurred on reaching a consensus on these issues. The Paris Agreement is silent on these vital issues except a mention of parties' responsibility to consider 'human rights' while addressing climate change (Preamble, para 8).

In view of the preceding, this chapter argues that reaching an equitable agreement on distribution of emission rights for current and future development in the PVCs hinges on solving the intractable issue of responsibility arising from the disproportionate share of historical emissions produced by the developed countries. These emissions enabled the industrial countries, many of which were colonial powers with access to vast natural resources, to advance their economic growth and standard of living. Therefore, they have to lead in reaching agreement (UNFCCC Article 3.1), absence of which will continue giving the major emitters from developing countries the space for businessasusual in their GHG emissions. Though the ethical and moral considerations regarding the distributional equity overwhelmingly support the cause of the PVCs, in global diplomacy steeped still in promoting power and material interests of nation-states, those ethical considerations donot go far. Therefore, discussions in this chapter step further into how the ethical and policy considerations can be translated into reality under the agreed framing of the climate regime, that is, neoliberalism and market solution.

For that purpose, next section of the chapter lays down the ethical strands and policy proposals regarding the distributional equity related to emissions rights and points out which approaches are ethically and practically fair for the PVCs. Section three provides an exposé of how the PVCs are caught in a triple bind of climate change impacts, unfavorable market-based mitigation approaches, and utter lack of adaptation finance. This latter is the most vital and immediate concern for the PVCs. Section four discusses how climate change infringes on the human security and development rights of the PVC citizens.

Section five incorporates the cardinal principles of market solution, that is, the polluter pays principle (PPP), based on the no harm rule and state responsibility. These principles are implicitly recognized in the Climate Convention[2], and the newly agreed UNFCCC agenda of loss and damage, together with mitigation and adaptation. This new agenda has the potential of germinating the mechanisms of liability and compensation. Finally, a Conclusion sums up the whole discussion.

Policy Proposals and Principles of Ethics Regarding Distribution of Emission Rights, or Responsibility for Emissions Reduction

Besides the discourse on principles of ethics regarding sharing of emission rights within the absorptive capacity of the atmospheric sink, four proposals that were floated in the UNFCCC negotiations stand out for brief discussion here. They are grandfathering, carbon intensity, global per capita emissions, and historical responsibility (Roberts and Parks, 2007).The grandfathering approach is based on the idea that the industrial nations that developed early through GHG emissions will commit resources for reducing their emissions relative to a base year. The Kyoto Protocol of 1997 was based on this approach, in which the industrial countries assumed a binding commitment to reduce GHG emissions on average bymore than 5 percent by 2008–12, compared to the base year of 1990. The EU as a group assumed responsibility for reducing emissions by 8 percent, but it had the application of the principle of equity and common but differential responsibility based on respective capabilities (CBDR + RC) within its member states, where the richest countries, such as Germany and the United Kingdom,committed to reducing more than15 percent, while allowing less developed members such as Portugal and Greece to increase their emissions. The United States never joined the Protocol on the ground that major developing countries such as China and India have been exempted from binding commitments. But this grandfathering approach did not succeed in the hands of some non-EU Kyoto parties, such as Canada, Japan, and Russia, have either opted out of the Kyoto process or did not join its second commitment period, which began in 2013.

The carbon intensity approach, initially advocated by the World Resources Institute (WRI), was supported by the U.S. administration of George W. Bush. It calls for voluntary improvement in energy efficiency, so that over time carbon

[2] Article 3.1 of the UNFCCC provides for the principle of equity and common but differentiated responsibility based on respective capabilities; in view of recognition of the largest share of historical and still continuing emissions originating in the developed countries and lower per capita emissions in developing countries, the former has been assigned the lead role in mitigation (Article 3.1), and in providing financial support for adaptation in the PVCs (Article 4.4); thus this provision implicitly combines historical responsibility (which is an implicit recognition of the PPP) and response-ability/capability. But the industrial countries focus on the ability part, while the developing countries focus more on the responsibility aspect of the provision.

emissions per unit of GDP decrease. As a win–win proposition, this approach supports economic growth with reducing carbon intensity. Industrial countries have better opportunities for applying this proposal, while developing countries are also interested for efficiency reasons. In this approach, developed countries can help developing ones by supplying cleaner technology (Article 4.7 of the Convention). Interestingly, beginning with the UNFCCC Conference of Parties 15 (COP15) in Copenhagen in 2009,[3]major emitters from developing countries such as China, India, Mexico, Brazil, South Africa, and Indonesia have committed themselves to reducing their carbon intensity at varying levels. Even China in the March 2014 UNFCCC negotiations in Bonn proposed that it would do so by reducing carbon intensity under the Durban Platform, instead of committing themselves to binding reductions.

The other two proposals are supported mainly by the developing countries. Since atmospheric sink capacity is a global commons, according to this approach, each citizen on Earth has an equal right to an average for all to emit. So, industrial countries have to reduce their GHG emissions, and developing countries' increase over time will converge somewhere to an agreed average, allowed by the remaining carbon space. The problem with this approach is that developing countries with big populations such as China, India, or Indonesia obviously gain, with the huge potential of using extra allowances for emissions trading. This proposal also encourages population growth. In order to apply, this proposal at a minimum needed to agree on a cut-off date for specifying population numbers in each country, but it never gained traction in the UNFCCC negotiations.

The final, yet the most debated is the proposal of assuming historical responsibility for past emissions by the early industrializers of the world. The originator of this proposal is Brazil, which back in 1997 proposed this approach. Implicit in this proposal is the polluter pays principle. The argument is that a country's reduction of GHG emissions should depend on its contribution to rise in global temperature (UNFCCC, 1997) at least since the recording of climate data, that is, 1860. Since GHGs are a stock rather than a flow problem, CO_2 remains in the atmosphere for more than a century, and this antecedent deposition has caused a rise in temperature for which the old industrializers have to take responsibility. For obvious reasons, developing countries are strong advocates of this approach, arguing that the principle of CBDR implicitly recognizes this. Though the calculations of relative contributions are complex, Roberts and Parks (2007:147–150) estimate that since 1950, the sum of emissions from the wealthy nations amounts to nearly twice the tons of CO_2 of the middle income countries, and four times the cumulative emissions of the majority of world nations. So they argue (p.150) that "this is a

[3] COP15, attended by more than 120 world leaders, was supposed to work out a new emissions reduction regime, as stipulated in the Bali Action Plan, adopted at COP13 in 2007, but it was an utter failure.

highly contentious issue, but one that we believe must be considered if we are to address inequality and climate change. The polluter pays argument is that high-emitting nations, even if they did not know the danger of their behavior, still benefitted from it and should be held responsible for its impacts." In section five, I will turn to why and how this principle, which has evolved from an economic to an ethical and legal norm, should be applied globally.

Neither of these proposals was succeeding or being accepted by the UNFCCC parties. Only the EU remains committed, but it is not being effective because the group's share of emissions is too low to have a global impact. As a result, quite a number of hybrid proposals cropped up, combining elements from different proposals, such as a 'preference score' method (Bartsch and Muller, 2000); per capita emissions considering 'national circumstances' (Gupta and Bhandari, 1999; Torvanger and Godal, 2004); WRI's proposal that combines responsibility based on past and present emissions, carbon intensity, and ability to pay (Claussen and McNeilly, 1998); the 'triptych' proposal by scholars at the University of Utrecht in the Netherlands, which is already applied within the EU member states and accounts for national differences within the EU (Groenenberg et al., 2001); and the latest proposal by Müller and Mahadeva, 2013), which is based on a process like progressive taxation policy. The Paris Agreement expanded the principle of CBDR+RC with the addition of "in light of national circumstances." So, this gives some leeway to developing countries, particularly to the PVCs, in their mitigation responsibilities.

However, the age-old saying that 'beauty lies in the eyes of the beholder' continues to hold in the climate negotiations. Fairness is interpreted subjectively by the UNFCCC parties and in ways that suit their national perspectives. So Roberts and Parks (2007:150) argue for the need for moral compromise or a negotiated "justice settlement." Meyer and Roser (2006) sketch three basic principles of distributive justice – egalitarianism, sufficientarianism, and prioritarianism – and they critically apply them to distribution of emission rights (interpreted as benefits from emissions) for the currently living generations. They reach the conclusion that prioritarian standards are the most plausible that give more emission rights to developing countries, where many citizens do not fulfill the basic needs of life. So, there is a strong moral case for 'serving the subsistence needs rather than luxury needs' of citizens in the industrial countries. In terms of historical responsibility, they argue that high past emissions cannot be justified for a right to enjoy above-average current emissions. While discussing briefly the impacts of climate change, they propose fund transfers from industrial countries to developing countries to address their adaptation needs; further, they suggest that instead of direct financial support, industrial countries may allow larger shares of emission rights to the LDCs, which they can monetize through the mechanism of emissions trading. Their arguments appear fully supportive of the concerns of the PVCs, and the UNFCCC provisions exactly represent this spirit. However, this potential gain in enhanced emissions quotas for the PVCs is likely to be nullified if developed countries do not stop transferring emissions-intensive industrial production to these countries. The existing Kyoto Protocol–based framework isnotdesigned in terms of consumption, but of production-based emissions accounting. So this is a great source of emissions leakage, and there is an incipient movement against this in the European countries (Boitier,2012; Steininger et al., 2015). However, the Paris Agreement also appears to be silent on this point. In the decades to come, unless consumption-based emissions accounting by countries is agreed upon, this leakage will stand in the way of effective mitigation efforts across the globe.

As posited earlier, moral or ethical considerations donot go far in international diplomacy. But I will argue that the solution regarding distribution of responsibility for reduction of GHG emissions or sharing of emission rights for the remaining space can be found within the agreed framing of the existing regime, with application of neoliberal and market-based ideas. Meanwhile, let me turn to show how climate change affects the development rights of the PVC citizens.

PVCs Are Trapped in a Triple Bind of Increasing Climate Change Impacts, Unsupportive Global Mitigation Measures, and Lack of Adaptation Finance

Evidence of increasing climate impacts has been mounting, as shown by the successive reports of the Intergovernmental Panel on Climate Change (IPCC) since 1990,including the latest report by the Working Group 1 of the IPCC. Its special report (IPCC,2012: 5) on extreme weather events warns us that "a changing climate leads to changes in the frequency, intensity, spatial extent, duration and timing of extreme weather and climate events." Even normally staid institutions such as the International Energy Agency and the World Bank in their latest reports have included warnings of runway climate change, an increase of 4^0C to 6^0C, and its dire consequences (IEA, 2012; World Bank,2012). The latest U.S. Assessment Report on Climate Change, prepared by the White House, links GHG emissions to 'unambiguous' climate changes in the United States (Morello, 2013). The president of the World Bank, Jim Yong Kim, emphatically warned of the impending dangers of climate change (*ClimateWire*, 12 October 2012). The UN chief argued in his speech at the UNGA on 9 November 2012 that "extreme *weather* due to climate change is the new normal .... This may be an uncomfortable truth, but it is one we ignore at our peril" (Ki-Moon, 2012). Already the global community is witnessing increasing frequency, severity, and magnitude of floods, cyclones, storm surges, drought, and heat waves. The heat waves in

Europe and drought in Russia; the continued and spreading drought in the United States; super storm Sandy that hit the northeastern coast of the United States in October 2012; floods and storm surges in the Bay of Bengal basin, frequent floods in China and Europe are considered examples of erratic behavior of the climate system.

Besides, climate change has a ratchet effect – existing stock of greenhouse gases is already warming up the earth and will continue to do so with more and more addition of GHGs, making needed actions more and more stringent. This phenomenon has temporal and spatial dimensions as well. The PVCs are concerned more with the spatial dimension of climate change, which inflicts multiple inequities in its impact distribution. As IPCC findings (2007b, 2012) show, African countries, SIDS, and low-lying coastal countries like Bangladesh will be most impacted from climate change and its extreme events. This is due both to their geographical locations and to other socioeconomic stresses. During the period 1980–2013, the LDCs suffered 51 percent of deaths from climate-related disasters – 4.7 times the global average; taking just January 2010 to July 2013, the figure rises to 67 percent, 5.5 times the global average (Ciplet et al.,2013).

Thus, Vanderheiden (2008) cogently argues that in both absolute and relative terms the poor lose from climate change in multiple ways: 1) climate change already takes an unjust global distribution and exacerbates its undeserved inequality to the detriment of the least advantaged; 2) while climate change results from voluntary acts and choices, those responsible suffer least, while those least causally responsible suffer most; and 3) those acts largely responsible for climate change – burning of fossil fuel and deforestation – simultaneously benefit the already affluent and harm the poor, in effect transferring welfare from the poor to the rich, thereby increasing disparities. Another aspect of this disparity lies in the fact that while agriculture contributes just 2–4 percent in industrial countries' GDP, it contributes about 18–60 percent in the LDC economies (World Bank, 2010), with more than two-thirds of the labor force depending on agriculture in these countries for their livelihood, directly or indirectly (Stern, 2009). Thus, IPCC (1995) and some other research (Stern, 2009) rightly projected that consequences of climate impacts are likely to "aggravate existing disparities" by worsening conditions and opportunities of the poor.

The impacts are already contributing to undoing the development efforts both of the past and of thepresent in these countries. Already a decade ago, a multidonor Report on Poverty and Climate Change (World Bank et al., 2003) rightly acknowledged that "climate change is a serious risk to poverty reduction and threatens to undo decades of development efforts" (see also Davidson et al., 2003). Other reports are even more alarming. The Human Impact Report of 2009 by the Global Humanitarian Forum led by K. Annan states that climate change is already killing 315,000 people a year and will have a severe impact on 600 million, almost 10 percent of the global population within twenty years (GHF,2009).

IPCC (2007a) projects that crop production in South Asia is likely to witness a drop of about 30 percent. This is likely to cause great human suffering globally, as 2.6 billion people now depend on agriculture for their livelihoods (Millennium Ecosystem Assessment, 2005). South Asia is the region that will bear the brunt of this impact, as more than 600 million people – more than half of the world's total poor – live between the foothills of the Himalayas of the North down to the Bay of Bengal in the South. As is known, the Himalayas are a lifeline to some 1.5 billion people living directly in the floodplains of its many rivers. The region's long and heavily settled coastlines along the Ganges–Brahmaputra–Mekong mega-delta are gravely threatened by sea-level rise.

Some empirical studies in different regions are showing the correlation between a rise of 1°C and loss in agricultural productivity or national revenue (Kurukulasuriya and Mendelsohn, 2008; Seo and Mendelsohn, 2008; Wang et al., 2008). Just a small decline in world output of wheat in the face of growing demand doubled its price in 2007. Several studies show that in a moderate climate scenario, the poorest countries could suffer almost two-thirds of global damages from climate change (Mendelsohn et al., 2006). The World Bank Report (2010) shows that developing countries will suffer 80 percent of global damage; with a severe climate change scenario, the poorest 25 percentof the countries of the worldcould suffer almost 50 percent of the damages (ibid.).

Some scholars argue that developing countries, particularly the LDCs, are likely to be affected less by direct climate change impacts and more by policy responses to address climate change (Sachs and Someshwar, 2012). Direct impacts result from bad harvests due to, for example, drought in major crop producing countries and the impact on price increases (Abbot et al., 2008, 2011; Wright, 2011). Food prices again increased in 2010–11, even more than the pre-crisis levels; in summer 2012, prices particularly of maize and wheat increased as a result of drought in major producing countries, with an extremely negative effect on the LDCs, where 50–80 percent of family income is spent on basic food items (UNCTAD, 2012). Analysts argue that along with a number of other suspects, a major reason was rise in energy prices leading to a surge in demand for biofuels from maize and oil seeds (Heady and Fan, 2010). This generated much debate on impacts of biofuels on long-term food security, particularly in the developing countries. The production of bio-ethanol in the United States already consumes a quarter of the corn (maize) crop and production of bio-diesel in the European Union, where the European farm lobbies' success in capturing huge government subsidies for these products yielded little or no net reduction in carbon emissions (Fitzroy and Papyrakis, 2010: 24).So, the ethical debate about food for the poor vs. fuel for the rich is raging across the North–South divide.

A further example of mitigation efforts' negative impact on poor local communities is the adverse impact of the UNFCCC-approved program of reducing emissions from degradation and deforestation plus (REDD+) in areas of poor governance and uncertainty of access of the poor communities to those resources. So, here is an instance of community livelihoods versus a market-based project of emissions reduction. Barr et al. (2010) argue that inequitable benefit sharing of REDD payments could increase disparities in the forestry sector and displace and impoverish forest-dependent peoples. In fact, under the rules of the neo-liberal market, it is the powerful who always benefit from the exchanges. The glaring example are the CDM projects, under which investments from industrial countries went overwhelmingly to major developing economies, such as China, India, Brazil, and South Korea. Of the total of 4,329 projects as of mid-2012, only 43 projects are being hosted by the 48 LDCs (only 1 percent), and of the close to 1 billion certified emissions reductions (CERs) issued by the UNFCCC, only about 200,000 went to the LDCs, a mere 0.02 percent of the total (ADB, 2011).

The experience of double exposure of low-income countries through climate change and uneven globalization shows that major goals of the world's powerful nations (such as trade agreements) are at risk if climate adaptation is not supported. But adaptation funding is utterly discriminatory – only about 20 percent of fast start finance (FSF) (2010–12) has been diverted to adaptation (Ciplet et al., 2013). Industrial countries have paid only U.S.$0.9 billion to the LDC Fund to implement the national adaptation programs of action (NAPAs), the total costs of which are around $5 billion (Ciplet et al., 2013). Worse still, an Oxfam International report (2012) argues that only 33 percent of the FSF can be considered new; the rest was meeting the aid pledges made before COP15 in Copenhagen, and at most one-fourth was additional to existing aid commitments. Against this trend, Stern (2009: 14) argues for adaptation to be part of development, which is the best form of adaptation. For example, little extra funding for disaster preparedness and management shows strong returns – in China expenditure on flood control of $3billion is estimated to have returns of $12billion; in India, disaster programs in Andhra Pradesh have shown a benefit–cost ratio of 13:1 or more; in Vietnam, planting mangroves to protect from typhoons and storms yields a benefit–cost ratio of 50:1 (Stern, 2009). So Khan (2014) argues that unless the framing and legal basis of adaptation are strengthened, it is likely to remain discriminatory. Khan argues that like mitigation, adaptation (the need for which will continue to emanate from undersupply of mitigation) should be recognized as a global public good in an age of atmospheric commons. One positive element in the Paris Agreement is that it establishes an explicit link between adaptation and levels of mitigation.

Climate Change, Security, and Development Rights

The low-income countries and poor people around the world are becoming double losers from the simultaneous processes of climate change impacts and uneven financial globalization of today. The PVCs with their weak economies are bypassed in this process, whereas they are exposed to unequal ecological exchange; declining terms of trade, particularly for agricultural products (the mainstay of economy in many countries); declining foreign aid in real terms; unequal impact of financial crises; and other deleterious effects (Parks and Roberts, 2010; Bhattacharya and Dasgupta, 2010).Climate change–induced scarcity of life-support resource systems, such as the food chain and water, is likely to cause conflicts, particularly in the already resource-constrained and degraded areas. Sub-Saharan Africa is cited as the premier example in this category of risks, with continued drought, followed by South Asia, battered by extreme climate events, such as floods and cyclones, and these are predicted to cause political instability and conflicts in the regions (Podesta and Ogden, 2008). Darfur is regarded by the UN secretary general as the 'first climate war' (Ki-Moon, 2007). Sachs (2006) argues that climate change is leading to warfare in Ethiopia and Sudan, which peacekeepers, sanctions, and humanitarian aid are not going to stop; instead, he suggests cutting emissions drastically by the developed world, while helping developing countries adapt. This kind of understanding requires sensitization of the policymakers, which is currently absent.

However, Podesta and Ogden (2008: 134) raise the difficulty of marrying the moral and security challenges posed by climate change:

> Ultimately, the threat of desensitization could prove one of the gravest threats of all, for the national security and foreign policy challenges posed by climate change are tightly interwoven with the moral challenge of helping those least responsible to cope with its effects. If the international community fails to meet either set of challenges, it will fail to meet them both.

Though there are different conceptualizations about the linkage of security and climate change, such as the lenses of environmental conflict and environmental security (cf. Paris, 2001; Barnett, 2009; CNA, 2007; Detraz, 2011), climate change is emerging as a human and global security issue (Khan, 2014:figure 1). It is a matter of a different lens with which to see the same kaleidoscope of direct threats or causal chains that generate threats either to state or to human security. The essence of this discourse is that impacts of accumulated development and adaptation deficit may spill over onto the front yards of industrial countries, particularly in the form of climate refugees (cf. Khan, 2014; Smith, 2007; Myers, 1993; Warner, 2010).Obviously, the defense departments of major countries of the world have already undertaken studies to explicate the linkage between climate change and security (cf. Aus DoD, 2009; US DoD, 2010; Holland and Vagg, 2013).

The climate conflict and refugee narrative may contribute to policies in different directions: one response could be tightening of state borders and militarization of development aid (Carlarne, 2009), while such a narrative may also remind nations of the need for strong mitigation and adequate adaptation in situ through mobilizing the pledged climate finance (Hartmann, 2010). In a similar vein, Carlarne (2009: 468) argues that "from an equity perspective, it is not yet possible to determine whether fears over climate-based security threats will promote global cooperation or protectionist policy-making – the latter could be devastating to global adaptation efforts." However, with climate impacts becoming more and more evident, the military might reshape their activity profile.

This linking of climate change with human, national, and global security makes the framing of a rights and justice lens to climate change really salient. It may be recalled that in his 1974 Hague Academy Lecture, the Nobel Prize winner Rene Cassin argued for extending the existing concepts of human rights protection to include the right to a healthy environment (cited in Picolotti and Taillant, 2003: 121). The Inuit petition of 2005 to the Inter-American Human Rights Commission for a hearing of their plight from climate change was the push for the union of climate change and human rights. The basic idea is that climate change impacts the livelihoods and amenities of life negatively such that it affects the enjoyment of life at a minimum level, particularly by the poor.

Some scholars, such as Rajamani (2010), Caney (2010), Shue (1993, 1999), Vanderheiden (2011), Moellendorf (2011) and others, argue that climate change impacts documented by the IPCC are already undermining and likely to undermine further the realization of a range of *protected*human rights, such as the rights to life, liberty, security, and development. Bell (2013) introduces the moral conception of human rights, as distinct from the human rights conceived in international and national laws. He summarizes three arguments for linking human rights to climate change impacts: first Vanderheiden's idea of having a human right to a stable climate; second, Caney's claims that climate change violates the basic human rights to life, health, livelihoods, and subsistence rights; and third, the rights to emissions, which he bifurcates into two categories: the right to equal per capita emissions and the right to subsistence emissions, which Shue (1993, 1999) advocates.

In fact, a compelling case for environmental rights follows from Henry Shue's pioneering work on "subsistence rights" and his differentiation between basic and non-basic rights (Shue 1993, 1999). Basic rights are understood by Shue as the threshold beneath which no one should be allowed to sink and so they constitute a person's minimum demand upon the rest of humanity. Shue argued that a stable climate is a basic right, as its destruction interferes with development rights of others (see also Raworth 2012). In a similar vein, Stern (2009) argues that emissions of GHGs interfere with the development rights of other people.

Vanderheiden (2008) offers a notion of climate justice based on both environmental and development rights; following Shue, he articulates development as a right, with basic environmental amenities including a stable climate system. The realization of the conceptions of capabilities and functioning of Sen (1999) is obviously dependent, among others, on environmental amenities.

However, Hayward (2007: 445) argues that emphasizing emissions as a right sounds negative. So she suggests from the climate justice perspective requesting an "equitable share of the planet's aggregate natural resources and environmental services that are available on a sustainable basis for human use." And this conception is consistent with the fundamental human right to an adequate environment. The concept of equal rights is particularly relevant for global common pool resources that exist outside the legal control of individuals or nation-states (Baer et al., 2000). For example, in terms of the notion of the common heritage of mankind, the UN Convention on the Law of the Sea (UNCLOS) codified common ownership of deep-sea resources for the benefit of all humanity. Baer et al. (2000) cite the precedent that governments have adopted egalitarian principles in allocating resource rights (e.g., the U.S. Public Trust Doctrine) even in cases when there were large and unequal pre-existing claims.

Caney (2010) forcefully argues that climate change violates at least three rights: a) the right not to be deprived of one's life (which is regarded as a peremptory right in any national and international legal regime), b) the right not to have others cause serious threats to one's health, and c) the right not to be deprived of one's means of subsistence. This no-harm rule is virtually accepted as a customary principle in the Western world. Together with the climate change impacts, the way mitigation measures are implemented also affects human rights: these are REDD+ initiatives, which are dislodging the communities of their customary rights to forests, and the production of bio-fuels, which negatively impactfood security and the price of basic food cereals (Bachram, 2004; McInerney-Lankford,2009; Sachs and Sameshwar, 2012).

Sen (2004: 316) refers to the U.S. Declaration of Independence of 1776, which apparently took it to be "self-evident" that everyone is "endowed by their Creator with certain inalienable rights," and thirteen years later, the French declaration of "the rights of man" asserted that "men are born and remain free and equal in rights" (Sen, 2004). These basic rights are extra-territorial in nature, recognized universally through the adoption of a thick layer of rights conventions. So, Sen is his work *The Idea of Justice* (2010) argues against the Rawlsian liberal notion that rights and social justice are territoriallybounded, that in an interdependent and globalized world, the notions of justice and fairness extend beyond a country's border. And of all these rights, basic, material rights have precedence over non-material and political rights (Sen, 1999).

Thus, a human rights lens to climate change, in both its prevention (mitigation) and adaptation aspects, is strongly evident in the scientific literature, in advocacy circles, and in climate negotiations. This "climate justice frame" is gaining a growing number of adherents, and their voices are growing louder (e.g., Climate Justice Now!, 2012; Amnesty International, 2009; Chawla, 2009). This linking of human rights and climate change as bandwagoning (Nicholson and Chong, 2011) is viewed to be enhancing the legitimacy of both regimes. Nicholson and Chong (2011: 126) argue that deploying in climate change a "human rights framework as an analytical device does two things: First, it provides a set of basic moral principles that place power inequities and injustice at the centre of the climate change debate. Second, it provides moral authority for calls to action."

Shue (1992) in a seminal contribution titled "The Unavoidability of Justice" argues that questions of justice are not external to international negotiations on climate change on three grounds: a) 'background injustice' is not lost sight of by the parties involved in negotiations, which overtime gave rise towhat some others call 'principled beliefs' (Roberts and Parks, 2007); b) the harm caused by the rich nations, though unintentional, is the subject of negotiation and cooperation; and c) avoiding the issue of justice would ultimately condemn the poor nations to sacrificing their 'vital' interests, that is, survival interests, in order for the rich nations to avoid sacrificing their 'trivial' interests. Shue has rightly said, "If one is profiting from injustice, it is hardly going to be in one's interest to pursue justice" (1992: 376).

Some scholars have devised frameworks as to how to translate the human rights–based approach to climate change. Baer et al. (2009) argue that their greenhouse development rights is a framework based on universal human rights compatible with but not identical tothe egalitarian principles; it is a class-based rather than a nation-based approach to economic justice. It considers equality both within and beyond nation-states, and national obligations are based on the exemption of the poor under a 'development threshold' from global burdens. This development threshold has been fixed at twenty dollars/day income, and those who are above this threshold in either the developed or the developing world are obligated to share the costs of addressing the climate change problem. On the basis of such a framing, they have calculated the capacity and responsibility index. This proposal implicates the rich in the global South, too, as sharing responsibility, but their share is much smaller than that of the industrial countries.

This justice and rights framing is gradually gaining traction in the UNFCCC process. This is reflected in the Preamble of the Cancun Agreements adopted at COP16 in Cancun in December 2010, which acknowledges the applicability of a human-rights framing for climate change deliberations, emphasizing the need that all parties "in all climate change–related actions, fully respect human rights"

(Cancun Agreements, Preamble, para 8). The Preamble of the Paris Agreement also does so. This discourse has entered into formal climate negotiations as well. Rajamani (2010), in her analysis of submissions of UNFCCC parties from 2008 to 2010, found that Argentina, Bolivia and Chile, Thailand, Iceland, and the LDCs explicitly argued for the relevance of a human rights approach. They proposed the insertion of a human rights perspective in the negotiation texts since the 2007 Bali COP13.

Though in climate regime texts the concept of justice is not directly specified, the terms and provisions in the convention relate to the conception of justice (some paragraphs in the Preamble, Articles 3.1, 3.4, 4.3, 4.4, etc.). Though not direct, those provisions have an implied meaning of justice (Okereke, 2008). Besides, the principle of CBDR implicitly refers to the polluter pays principle. The provisions, such as 'the need for equitable and appropriate contributions' by industrial country parties 'to the global effort' of mitigation (Article 4.2a), 'the need for adequacy and predictability in the flow of funds and the importance of appropriate burden sharing among the developed country Parties' (Article 4.3), and that 'Annex II shall also assist the developing country Parties that are particularly vulnerable to the adverse effects of climate change in meeting costs of adaptation' (Article 4.4), are clear recognition and acceptance of industrial countries' responsibilities. All these provisions indicate an *implied* acceptance of compensation to the developing countries by the Annex II parties. Actually, if these provisions are implemented in their true meaning and spirit, global justice can be ensured both in mitigation and in adaptation. But that is not happening as a result of the insincerity of the powerful parties. In order to overcome this 'gaming', let me now turn to argue that application of the polluter pays principle (PPP), the cardinal principle of market systems, might become the key to realization of development rights by the PVCs.

A Differential Application of the PPP between Developing and Industrial Countries

This section focuses on how some teeth can be added to realize the rights of the PVCs to development against climate change. This obviously will entail an increase of their existing level of emissions, as well as in enhanced adaptation support. There are two widely used principles that can be brought to bear, the polluter pays principle and the issue of liability and compensation, based on the no-harm rule.

Historically, the idea of PPP for environmental harm is rooted in both Western and Eastern traditions. Luppiet al.(2012: 135) cites as a footnote *The Dialogue of Plato: The Laws* (1953), the celebrated passage by Plato: 'If anyone intentionally spoils the water of another ... let him not only pay for damages, but purify the stream or cistern which contains the water.' The celebrated Indian philosopher Kautiliya, who lived more or less at the same time as Plato, in 300 BC, in his *Arthashastra* (Study of Economics), prescribed financial penalties depending on levels of harms caused to the environment (Kangle 1986).It is clear that these sages conceived of the PPP for application in management of the local commons. Gradually, it was applied as an economic instrument for pollution control (Sanford, 1991; Smets, 1994).In the 1980s, regulations were deemed more desirable and efficient in environmental protection (WCED, 1987: 198–200, 319).However, some change was reflected in Agenda 21 adopted in Rio in 1992. The new call was for international cooperation in the use of economic instruments (Agenda 21, 1992: 252–54). Still this approach is somehow not gaining traction at the global level.

There are different rationales or interpretations as to the efficacy of the PPP, of which the following four can be cited as the most common (cf. Pearson, 1994; Parikh, 1993; Nash, 2000; Luppi et al., 2012): an efficiency argument, an equity argument, a judicial/legal argument, and a pedagogical argument. Cost internalization of negative externalities as its core meaning is meant for efficient allocation of resources. This is also called full cost pricing. The idea is that once the polluters are bound to internalize the costs, they will try to reduce those costs by reducing pollution. So, there is a built-in incentive for R & D for new technology. The judicial/legal interpretation of the PPP holds that states and local governments are jointly and severally liable for environmental damage caused by the private parties, allowing the public authorities to act in 'subrogation' against industrial polluters (Luppi et al., 2012). In addition, Nash (2000) argues that there is a pedagogical argument for this principle, both for the producers and for the consumers: both these groups are instilled with a sense of responsibility for the pollution load that they generate through production or consumption. Nash further argues that politicians also are likely to like it, since supporting the PPP puts them on the side of the voters, as the principle has an inherent social and public interest appeal. Then, in its equity interpretation, it is understood in terms of fair distribution of costs. All four interpretations are extremely important for international climate policy formulation.

Thanks to the extensive work of the OECD during the last two decades, the PPP has been transformed from an economic to a legal principle (OECD, 1992:9).EC Directive 84/631 (6 December 1984) on control of transborder shipment of hazardous waste illustrates the application of PPP. In Russia and some former Soviet bloc countries, this principle is applied as an instrument of environmental regulation (Glazyrina et al., 2006). PPP is also applied in many different forms and ways in many developing countries. Luppiet al. (2012)argue that by reframing the original rationale of the PPP, its application has been transformed into a government-pays regime, where governments assume the liability to pay immediate compensation to victims. This is kind of a subsidy to the private sector, which is in the development stage in these countries.

U.S. domestic law did not codify PPP, but it did have an influence on the development of the U.S. environmental law in the 1970s and 1980s. For example, certain provisions of the 1970 Clear Air Act and the 1977 Clean Water Act (CAA) require polluters to meet certain standards at their own expense. The Comprehensive Environmental Response, Compensation and Liability Act 1980 mandates the fulfilling of PPP by imposing liability for cleanup costs on the polluters. Under the 1990 amended CAA, the United States introduced trading in sulphur emissions, which is a variant of the PPP. The 'Superfund' legislation in the United States held that polluters are liable for cleanup costs of hazardous sites, even if dumped materials were not known at the time to be harmful (Brennan, 1993).

An IMF study using a broader concept of subsidies to include failures to impose taxes on pollution externality and failures to impose taxes on energy that are comparable to taxes in other goods (Narayanan, 2013) found that energy subsidy globally stands at a colossal $1.9 trillion, of which industrial countries account for about 40 percent of the total including taxes forgone; the U.S. subsidy alone amounts to $502 billion. The study argues that removal of subsidies can alone reduce 13 percent of global CO_2 emissions by 2050 and calls for imposing a price on negative externalities of using fossil fuels, which is application of the PPP. This huge amount of subsidy can be put against the renewable energy worldwide, which received six times less support than fossil fuels (News/World, 3 December 2012). Steve Kretzmann of Oil Change International, an advocacy group for clean energy, argues that "we need to stop funding the problem, and start funding the solution" (cited in ibid.). There was an understanding to eliminate fossil fuel subsidies at the G-8 Summit held in October 2009.

Young (2010) very cogently rationalizes the application of the PPP as a "progressive development" of the post-2012 climate regime: while industries pay for managing solid wastes, GHG emissions do not require full-cost accounting, and this presents a serious anomaly. The irony is that earthly garbage dumps are not free, but the atmospheric dump is treated as free! This is perhaps because it is a global commons, where major emitters with material and other elements of power think they can afford to globalize the cost, while enjoying the benefits at their country levels. So, freeriding remains the norm of major emitters.Having the PPP codified internationally would mean that polluters causing climate change would initiate reduction of GHGs and would have to pay those who suffered the impacts and would be forced to undertake expensive adaptation measures. In negotiations some countries and groups, including AOSIS, LDCs, Bangladesh, Pakistan, Switzerland, and Ghana,argued for the application of the PPP in emissions management and formakingit a guiding principle of the post-2012 climate regime (AOSIS, 2008; Maldives on behalf of LDCs, 2008).

This is the reason why the instrument of border tax adjustment on imports between those imposing a price on pollution and externality non-internalizing regimes is a hotlydebated topic among the WTO parties. Lord Stern argues that "countries that are on track to price their industrial carbon emissions, like China and those in Europe, should make it clear that they will eventually slap a border tariff on imports from countries that lack such a price, like the United States... This is not just protectionism. It's an argument about proper pricing of inputs. And if countries subsidize their exports by not pricing carbon, that's perfectly logical and sound reason for making border adjustments" (Chemnick, 2013). The Nobel Prize–winning trade economist Paul R. Krugman issued his own endorsement of the PPP, arguing that carbon taxes at the border are "a matter of levelling the playing field, not protectionism" (cited in Mattoo et al., 2009: 1). Along this line, another Nobel Prize–winning economist, J. Stiglitz went even further, arguing for internalizing the true cost of natural resources in his book *The Price of Inequality* (2012).

However, applying the PPP globally at this moment without any differentiation between the developed and developing countries would be problematic in terms of equity and justice, and hence its universal acceptability. Principle 23 of the 1972 Stockholm Declaration is pertinent here:

> Without prejudice to such criteria as may be agreed upon by the international community, or to standards which will have to be determined nationally, it will be essential in all cases to consider the systems of values prevailing in each country, and the extent of the applicability of standards which are valid for the most advanced countries but which may be inappropriate and of unwarranted social cost for the developing countries.

Thus this principle was kind of a precursor to the CBDR + RC, enshrined in the Rio Declaration as well as in the UNFCCC. This was the reason perhaps why the WTO adjudication over environmental cases overwhelmingly has given preference to economic, rather than environmental considerations. However, in recent times, the WTO is mulling over considering the border tax adjustments as a form of global application of the PPP (Mattoo et al., 2009).

Rajamani (2006: 214) argues that climate change is the only environmental agreement with an operational provision of the CBDR principle, where Article 3.1 is juxtaposed with the principle of leadership of industrial countries; the CBDR finds content in Principle 7 of the Rio Declaration, which clearly draws a link between industrial countries' culpability and enhanced responsibility. In fact, Brazil's proposal to apply PPP as compensation for historical emissions (see UNFCCC, 1997) was rejected by the Annex I countries during the Kyoto negotiations in 1997. There are several procedural problems in applications to account for historical emissions (Caney, 2010; Posner and Weisbach, 2010). Posner and Weisbach (2010) argue that the responsibility argument in PPP is backward looking, focusing on wrongful behavior of the past, when the wrongs were not understood. And many of those who emitted GHGs are no longer alive. This is true, but the argument for avoidance should not stop here. Caney (2010) and many others suggest that payment for emissions should be made at least since the time the harm was recognized. This means the Annex I countries should pay at least since the 1980s at the latest. Calculations by the MATCH research group (2007) show that moving the baseline year by a few decades does not dramatically shift levels of historical responsibility. For example, shifting the first year of counting emissions all the way from 1890 to 1990 decreases the contribution of OECD Europe from 14 to 11 percent of the world total.

Finally, Caney proposes as complementary to the PPP the 'ability-to-pay principle' (APP), which can take care of both emissions of past generations and legitimate emissions of disadvantaged countries and groups of people. He calls the latter poverty-sensitive PPP. A strict application of PPP also will affect major developing countries such as China and India, since PPP is not based on capability, but payment for using the ecosystem services of the atmosphere, specifically its limited sink capacity. While PPP is primarily a market principle, APP is a principle of justice. The equity part of PPP relates to the equitable distribution of the cost of mitigation. Along this line, Page (2012) argued for the application of the beneficiary pays principle (BPP), one version of which he calls the 'unjust enrichment' principle. Thus, in essence the APP and the BPP are not different. Ability to pay by some countries is the result of the benefits that they derived historically from burning fossil fuels for their development.

The model of Greenhouse Development Rights (Baer et al., 2009), which links climate change to a responsibility and capacity index, with a universal development threshold, appears more appropriate, in terms of justice and fit with the convention process (see also Müller et al., 2007).In a recently published work, Müller and Mahadeva (2013) propose a new framework (the Oxford Capability Measure, OCM) for measuring national differentiated economic capabilities (ATP) as an integral part of operationalization of the CBDR+RC.In their calculations, they use the analogous concepts of gross and net taxable income in a country; to illustrate the application of their methodology, they consider two examples: assessing the fairness of a given cost distribution and developing a (rule-based) 'graduation scheme' regarding obligations to pay. They conclude that "while an 'OCM-intensity of GDP'-based scheme would be best, one could use 'poverty intensity of GDP' as a second best surrogate" (Müller and Mahadeva, 2014: 2). In this approach, major emitters such as India as a low-income country do not have any obligation in costsharing.

PPP has several advantages in resource mobilization over domestic budgetary resources: public opposition is strong in allocating chunks of aid from domestic resources, the fund transfer will not be based on charity, PPP will satisfy new and additional criteria of climate finance, and there will be predictability, and, above all, prevention of harm through adequate mitigation is a lot cheaper than reactive or planned adaptation (Vanderheiden, 2011). Some scholars find the potential in the PPP for ensuring sustainable development both in the North and in the South, if applied in appropriate ways (Kettlewell, 1992).That polluters should pay the costs of dealing with their pollution reflects the most fundamental principles of justice and responsibility. It needs to be understood by the major polluters that to reach a fair outcome in assuming responsibility, the question of fair allocation of costs must be addressed. Meyer and Roser's (2006) ethics-based proposal for giving priority to the poor in emission rights or in adaptation funding fits this approach. Simon Caney, reflecting the sentiments of the PVC citizens, writes: "We cannot accept a situation in which there are such widespread and enormously harmful effects on the vulnerable of this world" (Caney 2005).

Thus, an efficient and equitable application of the PPP *ex-ante* can address the issue of historical responsibility. If not, the need for expost liability and compensation based on the no-harm principle and state responsibility may arise in the future. In the end, global application of the PPP warrants the fulfillment of a few conditions:a) For developing countries to apply PPP, industrial countries are required to make transfers of resources, financial and technological, so that the former can improve environmental standards in their production process; b) there must be agreement on the specific year from which to assume historical responsibility by the industrial countries for past GHG emissions; and c) a nation-state causing damage to another should bear the responsibility and pay compensation for it. The Paris Agreement (Article 8), however, forecloses the option of involving liability and compensation while dealing with the agenda of loss and damage. But it can be argued that for other parts of the agreement, such as those on mitigation or adaptation, the payment of compensation is not closed.

No-Harm Rule and State Responsibility

The principle of CBDR embodies commitments and rights of nations in environmental negotiations. Accordingly, UNFCCC provides a legal basis to claim support for damage prevention measures (both for adaptation and for mitigation) (Verheyen, 2005: 107).Rajamani (2006: 214) argues that two factors speak of 'culpability' of industrial countries: a) the convention contains language, though moderate, that lays blame on industrial countries, with the recognition (in the Preamble) of the "largest share of historical and current emissions originating in industrial countries," and b) the recognized need for them to take 'immediate action'; though the link is not absolute, still itis a recognition of culpability. According to her, no other international agreement contains even in the Preamble such an acknowledgement (ibid.).

Thus, in terms of interpretations of the provisions, it can be deduced that the climate regime provides a partial response to state responsibility through stipulations, which suffer from ambiguity and lack of teeth for enforcement. Paragraph 8 of the UNFCCC Preamble echoes the no-harm rule enshrined in Principle 21 of the Stockholm Declaration of 1972 and Principle 2 of the Rio Declaration of 1992, which urge states not to cause damage to areas beyond national jurisdictions. Article 194.2 of the UN Convention on the Law of the Sea (UNCLOS) implicitly bars states from creating unlimited emissions, and in case of failure, Article 235 obligates states to assume responsibility and liability. But this only relates to marine pollution. Further, Principle 22 of the Stockholm Declaration and Principle 13 of the Rio Declaration talk of cooperation for development of international and national laws regarding liability and compensation for the victims of environmental damage.

Many papers have been written specifically about moral responsibility (cf. Vanderheiden, 2008, 2011; Duss-Otterstrom and Jagers, 2012; Okereke, 2008), but it has been argued that they do not translate that responsibility into action.In environmental law, the 'no-harm rule' has become a customary international law. This rule is extremely important in climate change. The International Law Commission's "Draft Articles on State Responsibility for Internationally Wrongful Acts" provides that states are obliged to compensate for damage caused beyond one's border. Climate change has already crossed the threshold of tolerance. As Linklater (1998: 84) puts it: "Condemnation of transnational harm requires a commitment to regard insiders and outsiders as moral equals. Transnational harm provides one of the strongest reasons for widening the boundaries of moral and political communities to engage outsiders in dialogue about matters which affect their vital interests." This line of arguments aligns with Sen'sconcept of justice, as opposed to Rawlsian territorially-bounded theories of social justice (Sen, 2010).

Further, the 'do no harm' principle is a sufficient justification for a stringent mitigation policy as well as for adequate financing for adaptation. As a universally held value, a right not to be harmed implies a duty not to impose risk of harm on others, or pay compensation in case harm occurs. The principle can be viewed as a hierarchy of harm, in which death, injury, and physical suffering should take priority over economic costs or deprivation of property (Baer and Sagar, 2010).

The important legal case usually cited in the no-harm rule is the Trail Smelter Arbitration, in which sulphur dioxide emissions, lead dust, and other metallic compounds from a Canadian zinc and lead smelter had caused damages across the Columbia River border in the United States. In 1941 the court took cognizance of state responsibility and determined that the government of Canada had to compensate for damage that the smelter caused to land along the Columbia River valley in the United States(Kiss and Shelton, 2007). The principle of the 'no-harm rule' was also applied in the Lake Lanoux case and in the case concerning the Gabcikovo-Nagymaros Dam (Hungary v. Slovakia).These and other cases, including nuclear tests, confirmed the legal validity of the no-harm rule (Verheyen, 2005), and the provision of compensation.

There are strict liability provisions in many international conventions and treaties and the climate regime can learn from them, for example, concerning the sharing of responsibility under joint liability and contribution to the funds by stakeholders. However, most of the liability provisions discussed concern accidents or wrongful acts. Some scholars differentiate responsibility from a wrongful act and responsibility from a legal act (Kiss and Shelton, 2007).Greenhouse gas emission results from a legal act, but it harms others, particularly those who are the least emitters. This is the reason why the implementation of PPP is most appropriate among states, to take care of the responsibility of preventing harm by agents, or pay for damages caused, be it by private or state parties (ibid.). However, the GHG emissions cannot be considered legally valid at least from the moment when a state enters into legally binding commitments to reduce emissions. Even if there is any damage from a lawful act, the state concerned has to take responsibility for that damage. Further, whether an act is wrongful or not is determined by reference to due diligence (Birnie et al., 2009).Due diligence is argued to have included the following elements: a) the opportunity to act or prevent, b) foreseeability or knowledge that a certain activity could lead to transboundary damage, and c) proportionality of measures required to prevent harm or minimize risk (Schwarte and Byrne, 2010).Breach of these elements amounts to a wrongful act (Verheyen and Roderick, 2008). Some scholars argue for strict liability or fault-based liability, in which private agents or states are strictly liable for environmental harm, whether they knew of the impacts a priori or not (Posner and Weisbach, 2010).

Farber (2007) argues that potential damage from climate change is a legally cognizable injury, and he cites the case of *Massachusetts v. EPA*, which holds that plaintiffs did have standing to sue and the EPA does have the authority to regulate GHG emissions from motor vehicles. Farber (2007) also cites Section 601 of the U.S. Restatement (Third) of Foreign Relations Law (1987), which specifies that a state is obligated to take steps in not harming other states. Therefore, the 'no-harm rule' is regarded as the basis of an enforceable obligation by states to mitigate climate change and assist for adaptation (Verheyen, 2005).

However, the no-harm rule remainsmore of a soft law beyond Europe and the United States. Therefore, compliance with this rule through the application of the

PPP by states can achieve what the PVCs need – their right to development, in terms of both enjoying benefits from enhanced but legitimate emissions as well as from financial support for adaptation. However, the irony is that market mechanisms of which the PPP is the cardinal one are applied very selectively, based on narrow national interests, rather than the way Adam Smith, the guru of classical economics, suggested, with his accompanying moral arguments (1759). It is a contradiction that while the global leaders of the neo-liberal market system advocate liberalization and pressure developing countries to liberalize their economies and integrate into the globalized economy, their own economies donot introduce emissions trading or cap-and-trade systems or stand in the way of globalizing emissions trading. So, here is a double standard, in both ethical terms and policy approaches. However, when the legislative and executive branches are failing, the courts are stepping into the scene, with hundreds of climate litigation cases. At the national level, there are already some successes (Burns and Osofsky, 2009).

Conclusion

The preceding discussions show that the PVCs are especially disadvantaged by climate change in their development pursuits. Climate change with its most severe impacts on these countries tends to undo their development gains. As a consequence, disparity in development between the major emitters and the PVCs as the nano-emitters is likely to exacerbate. Human and national security and rights to develop are being challenged by climate impacts. The market-based solution is being applied under the existing climate regime in a truncated way, that is, through CDM and REDD+ in developing countries, and this bypasses the PVCs. Market outcome favors only those with market power. Some market power could have been given to the PVCs with higher emission quotas within the science-determined remaining capacity of the atmospheric sink, which could have been traded internationally by these countries. As argued before, the offshoring of emission-intensive industries by the industrial countries warrants calculating a country's emission level not on the basis of its production within its territory, but on the basis of consumption of goods and services from domestic production and imports. Without this approach, higher emission quotas in the PVCs would lose their utility, and with this leakage mitigation will not be effective. Together, because adaptation to the already manifesting climate impacts in the PVCs is their immediate priority, it should be financed in accordance with the agreed provisions of new and additional, adequate, and predictable funding (UNFCCC Article 4.3).

However, the crux of intractability continues to lie in the unresolved issue of responsibility of early industrializers for their historical emissions. This issue holds the UNFCCC parties back in operationalizing an effective universal regime even after adoption of the Paris Agreement, which is weak in terms of ambitious mitigation goals, a very weak non-penalty-based compliance mechanism, and financing, particularly for adaptation. So the acrimony is likely to continue over the issue of distributional equity in burdensharing for reduction of GHG emissions under the envisioned bottom-up approach, or allowing emissions space for late-comers in the industrialization process. Still there is no evidence of either power of example (in mitigation) or power of the purse (both for mitigation and for adaptation) from the early industrializers (Khan, 2014). This is likely to give the major emitters from developing countries space for businessasusual in their mitigation policy. Therefore, a base year needs to be agreed from which responsibility for emissions can be calculated, and this could be a period at least when the harmful effects of GHG emissions became known to the global community.

This chapter has shown that ethical principles and moral considerations, such as prioritarianism, climate justice, and assisting the disadvantagedfully support the PVC predicaments. But climate diplomacy hovers for years and years between the known causers of and known sufferers from the problem. Utter lack of adaptation finance as a consequence of non-compliance with pledges by the wealthy countries is the evidence that mere ethical and moral prodding does not work. Therefore, this chapter has argued for application of the polluter pays principle and no-harm rule, implied in the agreed provisions of the UNFCCC. The regime is based on neo-liberal, market principles, but when it comes to global application, major emitters continue to balk. However, there is already a global movement gradually gaining strength to impose carbon pricing. Now we have to look forward to whether the Paris Agreement will take force within the stipulated timeframe and how its provisions will be operationalized and implemented in the next decade and beyond.

References

Abbot, P. C., Hurt, C., and Tyner, W. E.(2008). *What's Driving Food Prices?* Farm Foundation. URL: www.farmfoundation.org/news/articlefiles/404-FINAL%20WDFP%20Report%207–28-08.pdf.

Abbot, P. C., Hurt, C., and Tyner, W. E.(2011). *What's driving food prices in 2011?* Farm Foundation. URL: www.farmfoundation.org/news/articlefiles/1742-FoodPrices_web.pdf. Accessed 4 January 2012.

ADB (Asian Develooment Bank). (2011). Workshop at the ADB Headquarters on: Enhancing regional distribution of CDM projects in Asia and the Pacific, 18-20 July. URL: www.iges.or.jp/en/cdm/pdf/regional/20120718/D2/S42_KaoruOgino.pdf. Accessed 12 March 2012.

Amnesty International (2009). *Upholding Human Rights while Confronting Climate Change*, London: Amnesty International Online. URL: www://amnesty.org/en/appeals-for-action/upholding-human-rights-while-confronting-climate-change. Accessed 22 October 2012.

Alliance of Small Island States (AOSIS) (2008). *AOSIS Input into the Assembly Paper on Financing*, Bonn: UNFCCC, FCCC/AWLCA/2008/Misc.5/Add.2 (part 1), 10 December, 17–23.

Department of Defense Australia (AusDoD)(2009). *Forces 2030: Defense White Paper*. Canberra: DoD.

Bachram, H. (2004). Climate Fraud and Carbon Colonialism: The New Trade in Greenhouse Gases. *Capitalism Nature Socialism*, 15(4), 5–20.

Baer, P., Harte, J., Haya, B., Herzog, A.V., Holdren, J.,Hultman, N. E., Kammen, D. M., Norgaard, R. B., and Raymond, L. (2000). Climate Change: Equity and Greenhouse Gas Responsibility. *Science*, 289(5488), 2287.

Baer, P., Kartha, S., Athanasiou, T., and Kemp-Benedict, E. (2009). The Greenhouse Development Rights Framework: Drawing Attention to Inequality within Nations in the Regional Climate Policy Debate. *Developoment and Change*, 40(6), 1121–38.

Baer, P., and Sagar, A. (2010). Ethics, Rights and Responsibilities. In *Climate Change Science and Policy*, ed. S. H. Scheneider, A. Rosencranz, M. D. Mastrandrea, and K. Kuntz-Duriseti.Washington, DC: Island Press, pp. 262–69.

Barnett, J. (2009). Human Rights and Vulnerability to Climate Change. In *Human Rights and Climate Change*, ed. S. Humphreys. Cambridge: Cambridge University Press, pp. 257–71.

Barr, C., Dermawan, A., Purnomo, H., and Komarudin, H. (2010). *Readiness for REDD: Financial Governance and lessons from Indonesia's Reforestation Fund*. Bogor: CIFOR Risk Management Projects in Practice.

Bartsch, U., and B. Müller. (2000). *Fossil Fuels in a Changing Climate*, Oxford: Oxford University Press.

Bell, D. (2013). Climate Change and Human Rights.*WIREs Climate Change*,4, 159–70.

Bhattacharya, D., and Dasgupta, S. (2010). Global Financial and Economic Crisis: Exploring the Resilience of the Least Developed Countries. *Journal of International Development*, 24, 673–785.

Birnie, P., Boyle, A., and Redgewelle, C. (2009). *International Law and the Environment*, 3rd ed. Oxford: Oxford University Press.

Boitier, B. (2012). CO_2 emissions production-based accounting vs. consumption: Insights from the WIOD databases. URL: www.wiod.org/conferences/groningen/paper_Boitier.pdf. Accessed on April 14, 2016.

Brennan, T. A. (1993). Environmental Torts. *Vanderbilt Law Review*, 46, 1–73.

Burns, W., and Osofsky, H. (2009). *Adjucating Climate Change: State, National and International Approaches*. Cambridge: Cambridge University Press.

Caney, S. (2005). Cosmopolitan Justice, Responsibility and Climate Change. *Leiden Journal of International Law*, **18**: 747–75.

Caney, S. (2006). Cosmoploitan Justice, Rights and Climate Change. *Canadian Journal of Law and Jurisprudence*, 18, 747–75.

Caney, S. (2010). Climate Change, Human Rights and Moral Thresholds. In *Human Rights and Climate Change*, ed. S. Humprey. Cambridge: Cambridge University Press, pp. 69–90.

Carlarne, C. (2009). Risky Business: The Ups and Downs of Mixing Economics, Security and Climate Change. *Melbourne Journal of International Law*, 10, 439–62.

Chawla, A. (2009). Climate Justice Movements Gather Strength. In *State of the World 2009: Into a Warming World*, ed. Worldwatch Institute. New York: W.W. Norton, pp. 119–21.

Chemnick, J. (2013). Americans want coasts, not feds, to bear the costs of sea level rise. *Greentewire*, March 29.

Ciplet, D., Roberts, T., Durand, A., Kopin, D., Santiago, O., Madden, K., Purdom, S., Chandani, A., Ousman, P., and Jarju, P. (2013). *A Burden to Share? Addressing Unequal Climate Impacts in the Least Developed Countries*. Briefing. Providence, RI: International Institute for Environment and Development (IIED) and Brown University.

Clausen, E., and McNeilly, L. (1998). *Equity and Global Climate Change: The Complext Elements of Fairness*.Arlington, VA: Pew Center on Climate Change.

Climate Justice Now! (2012). Climate Justice Now! "A network of organizations and movements from across the globe committed to the fight for social, ecological and gender justice," URL: www.climate-justice-now.org/. Accessed September 10, 2012.

Center for Naval Analysis (CNA)(2007). *National Security and the Threat of Climate Change*. Alexandria, VA: CAN Corporation.

Davidson, O., Halshnaes, K., Huq, S., Kok, M., Menz, B., Sokona, Y., and Verhagen, J. (2003). The Development and Climate Nexus: The Case of Sub-Saharan Africa. *Climate Policy*, 3(1), 97–113.

Detraz, N. (2011). Threats or Vulnerabilities? Assessing the Link between Climate Change and Security. *Global Environmental Politics*, 11(3), 104–18.

Duss-Otterstrom, G., and Jaggers, S.C. (2012). Identifying Burdens with Copying with Climate Change: A Typology of Duties of Climate Justice. *Global Environmental Change*, 22, 746–53.

Farber,D. A. (2007). Basic Compensation for Victims of Climate Change. *University of Pennsylvania Law Review*, 155, 1605–56.

Fitzroy, F. R., and Papyrakis, E. (2010). *An Introduction to Climate Change: Economics and Policy*.London: Earthscan.

Gardiner, S. M., and Hartzell-Nichols, L. (2012). Ethics and Global Climate Change. *Nature Education Knowledge*, **3**(10), 5.

Glazyrina, I., Glazyrin, V., and Vinnichenko, S. (2006). The Polluter Pays Principle and Potential Conflicts in Society. *Ecological Economics*, 59, 324–30.

Global Humanitarian Forum (GHF), (2009). *The Anatomy of a Silent Crisis: Human Impact Report Climate Change*, Geneva, GHF.

Groenenberg, H., Phylipsen, D., and Blok, K. (2001). Differentiating Commitments World Wide: Global Differentiation of GHG Emissions Reductions Based on the Triptych Approach – a Preliminary Assessment. *Energy Policy*, **29**, 1007–30.

Gupta, S., and Bhandari, P. (1999). An Effective Allocation Criterion for CO_2 Emissions. *Energy Policy*, **27**, 727–36.

Hartmann, B. (2010). Rethinking Climate Refugees and Climate Conflict: Rhetoric, Reality and the Politics of Policy Discourse. *Journal of International Development*, **22**, 233–46.

Hayword,T. (2007). Human Rights Versus Emissions Rights: Climate Justice and the Equitable Distribution of Ecological Space. *Ethics and International Affairs*, **21**(4), 431–450.

Heady, D., and Fan, S. (2010). Reflections in on the Global Food Crisis: How Did It Happen? How Has It Hurt? and How Can We Prevent the Next One? *IFPRI Research Monograph* **165**.

Heyward, M. (2007). Equity and International Climate Change Negotiations: A Matter of Perspective. *Climate Policy*, 7(6), 518–34.

Holland, A., and Vagg, X. (2013). *The Global Security Defense Index on Climate Change: Preliminary Results*. American Security Project. Washington, DC.

International Energy Agency (IEA) (2012). *Energy Outlooks 2012*. Paris.

Intergovernmental Panel on Climate Change (IPCC) (1995). *Climate Change 1995: Economic and Social Dimensions of Climate Change: Contribution of Working Group III to the Second Assessment Report of the Intergovernmental Panel on Climate Change*, ed. J.P. Bruce, H. Lee, and E.F. Haites. Cambridge: Cambridge University Press.

Intergovernmental Panel on Climate Change (IPCC) (2007a). *Climate Change 2007: The Physical Science Basis, Summary for Policy-Makers*. Geneva: WMO & UNEP.

Intergovernmental Panel on Climate Change (IPCC) (2007b). Summary for Policymakers. In *Climate Change 2007: Impact, Adaptation and Vulnerability: Contribution of the Working Group II to the Fourth Assessment Report of the Intergovenmental Panel on Climate Change*, ed.M.L. Parry, M.L Conziani, J.P. Palutikof, P.J. van der Linden, and C.E. Hanson. London: Oxford University Press, p. 2–5.

Intergovernmental Panel on Climate Change (IPCC) (2012). Special Report on Managing the Risks of Extreme Events and Disasters to Advance Climate Change Adaptation, Summary for Policymakers, ed. Field, C.B., Barros, V., Stocker, T., Dahe, Q., Dokken, D., Ebi, K., Mastrandrea, M., Mach, K., Plattner, G-K., et al.

Kangle, R.P. (1986). *Kautiliya Arthasastra* (Part II, English translation), Delhi: Motilal Banarasidass.

Khan, M. R. (2014). *Toward a Binding Climate Change Adaptation Regime: A Proposed Framework*. London: Routledge.

Kettlewell, U. (1992). The Answer to Global Pollution? A Critical Examination of the Problems and Potential of the Polluter-Pays-Principle. *Colorado Journal of Environmental Law and Policy*, 3, 492.

Ki-Moon, B. (2007). A Climate Culprit in Darfur, *The Washington Post*, June16. URL: www.washingtonpost.com/wp-dyn/content/article/2007/06/15/AR2007061501857.html. Accessed January 26, 2010.

Ki-Moon, B. (2012). Speech of the UN Secretary General ban Ki-Moon at the General Assembly on the Impact of Hurricane Sandy, November 9. URL: www.un.org/apps/news/infocus/sgspeeches/statments_full.asp?statID=1697#.V2Ts7hKM-1s. Accessed January 14, 2013.

Kiss, A., and Shelton, D. (2007). *International Environmental Law*. Leiden/Boston: Martinus Nijhoff.

Kurukulasuriya, P., and Mendelsohn, R. (2008), A Ricardian Analysis of the Impact of Climate Change on African Cropland.*African Journal of Agriculture and Resource Economics* **2**(1), 1–23.

La Rovere, E. L., de Macedo, L. V., and Baumert, K. A. (2002). The Brazilian Proposal on Relative Responsibility for Global Warming. In *Building on the Kyoto Protocol: Options for Protecting the Climate*, ed. K. Baumert. Washington, DC: World Resources Institute, pp. 157–71.

Linklater, A. (1998). *The Transformation of Political Community: Ethical Foundations of the Post-Westphalian Era*, Polity Press.

Luppiet, B., Parisi, F., and Rajagopalan, S. (2012). The Rise and Fall of the Polluter Pays Principle in Developing Countries. *International Review of Law and Economics*, 32, 135–44.

Maldives on behalf of the LDCs. 2008. International air passenger adaptation levy, Bonn: UNFCCC. FCCC/AWGLCA/2009/MISC.1, March 13, 2009, 59–60.

Modeling and Assessment of Contributions to Climate Change (MATCH) (2007). Summary report of the ad hoc group for the modeling and assessment of contributions to climate change (MATCH)'. URL: www/ match-info.net/. Accessed August 18, 2012.

Mattoo, A., Subramanian, A., van der Mensbrugghe, D., and He, J. (2009).Reconciling Climate Change and Trade Policy. *Center for Global Development Working Paper*, 189, 45.

McInerney-Lankford, S. (2009). Climate Change and Human Rights: An Introduction to Legal Issues. *Harvard Environmental Law Review*, 33, 431–37.

Mendelsohn, R., Dinar, A., and Williams, L. (2006). The Distributional Impact of Climate Change on Rich and Poor Countries. *Environment and Development Economics*, 11, 159–78.

Meyer, L. H., and Roser, D. (2006). Distributive Justice and Climate Change. *Analyse & Kritik*, 28, 223–49.

Michaelowa, A. (2001). Mitigation Versus Adaptation: The cpolitical economy of competition between climate policy strategies and the consequences for developing countries, HWWA Discussion Paper 153, Hamburg Institute of International Economics,

Millennium Ecosystem Assessment (2005). *Ecosystems and Human Well-Being*, Washington, DC.: Island Press.

Morello, L. (2013). Federal Report Links Man-Made Emissions to Unambigious Changes in US Life, Landscape. *Greenwire*, January 11.

Müller, B., Höhne, N., and Ellermann, C. (2007) *Differentiating (Historic) Responsibilities for Climate Change: Summary Report*.Oxford: Oxford Institute for Energy Studies.

Müller, B., and Mahadeva, L. (2013). *The Oxford Approach: Operationalizing "Respective Capabilities."* European Capacity Building Initiative. Oxford.

Müller, B., and Mahadeva, L. (2014). *The Oxford Approach Operationalizing the UNFCCC Principle of Respective Capabilities*, rev. 2nd ed. Oxford: European Capacity Building Initiative.

Moellendorf, D. (2011). A Normative Account of Dangerous Climate Change. *Climatic Change*, 108, 57–72.

Moellendorf, D. (2012). Climate Change and Global Justice. *WIREs Clim Change*, 3(2),131–43.

Myers, N. (1993). *Ultimate Security*. New York: W.W. Norton.

Narayanan, N. (2013). IMF Calls for a Carbon Tax, Claims Energy Subsidy Reform Would Make Large CO_2 Cuts. *Climate Wire*, March 28.

Nash, J.R. (2000). Too Much Market? Conflict between Tradable Pollution Allowances andthe Polluter Pays Principle. *Harvard Environmental Law Review*, 24(2000), 1–59.

News/World, December 3, 2012.

Nicholson, S., and Chong, D. (2011). Jumping on the Human Rights Bandwagon: How Rights-Based Linkages Can Refocus Climate. In *Global Environmental Politics*, 11(1), 121–36.

Okereke, C. (2008). *Global Justice and Neoliberal Environmental Governance: Ethics, Sustainable Development and International Co-Operation*. London: Routledge.

Organisation for Economic Co-operation and Development (OECD) (1992). *The Polluter-Pays Principle: OECD Analyses and Recommendations*. Paris: OECD.

Oxfam International (2012). The Climate 'Fiscal Cliff': An Evaluation of Fast Start Finance and Lessons for the Future.

Page, E. A. (2008). Distributing the Burdens of Climate Change. *Environmental Politics*,17(4), 556–75.

Page, E. A. (2012). Give It Up for Climate Change: A Defence of the Beneficiary Pays Principle. *International Theory*, 4(2), 300–330.

Paris, R. (2001). Human Security: Paradigm Shift or Hot Air? *International Security*, 26(2), 87–102.

Parks, B. C., and Roberts, J. T. (2010). Climate Change, Social Theory and Justice. *Theory, Culture & Society*, 27(2–3), 134–166.

Patz, J.A., Holloway, T., and Folley, J. (2005). Impact of Regional Climate Change on Human Health. *Nature*, 438, 310–317.

Pearson, C. S. (1994). Testing the System: GATT + PPP =?*Cornell International Law Journal*, 27(3), 553–575.

Parikh, K. S. (1993). The Polluter-Pays and User-Pays Principles for Developing Countries: Merits, Drawbacks and Feasibility. In *Fair Principles for Sustainable Development: Essays on Environmental Policy and Developing Countries*, ed. Edward Dommen. Hants, UK: Aldershot, 81–91.

Picoletti, R., and Taillant, J.D. (2003). *Linking Human Rights and the Environment*, Tuscan: The University of Arizona Press.

Plato (1953). *The Dialogues of Plato: The Laws*, 4th ed. Trans. Benjamin Jowett. Oxford: Clarendon Press.

Podesta, J., and Ogden, P. (2008). The Security Implications of Climate Change. *The Washington Quarterly*, 31(1), 115–138.

Posner, E. A., and Weisbach, D. (2010). *Climate Change Justice*. Princeton, NJ, and Oxford: Princeton University Press.

Rajamani, L. (2006). *Differential Treatment in International Environmental Law*. Oxford: Oxford University Press.

Rajamani, L. (2010). The Increasing Currency and Relevance of Rights-Based Perspectives in the International Negotiations on Climate Change. *Journal of Environmental Law*, 22(3), 391–429.

Raworth,K. (2012). A safe and just space for humanity: Can we live within the doughnut? Oxfam GROW programme discussion paper, Oxfam-GB, London. URL: www.oxfam.org/en/grow/policy/safe-and-justspace-humanity. Accessed August 31, 2012.

Roberts, J. T., and Parks, B. (2007). *Climate of Injustice: Global Inequity, North-South Politics and Climate Policy*. Cambridge, MA: The MIT Press.

Sachs, J. (2006), *The End of Poverty: Economic Possibilities for Our Time*. New York: Penguin Books.

Sachs, J., and Sameshwar, S. (2012). *Green Growth and Equity in the Context of Climate Change: Some Considerations*. Asian Development Bank Working Paper 371, July, Tokyo.

Sanford, G. (1991). The Polluter Pays Principle: From Economic Equity to Environmental Ethos. *Texas International Law Journal*, 26, 463.

Schwarte, C., and Byrne, R. (2010). International Climate Change Litigation and Negotiation Process. Working Paper, London: Foundation for International Environmental Law and Development (FIELD).

Sen, A. (1999). *Development as Freedom*.Oxford: Oxford University Press.

Sen, A. (2004). Elements of a Theory of Human Rights. *Philosophy and Public Affairs*,32 (4), 315–56.

Sen, A. (2010). *The Idea of Justice*. London: Penguin Books.

Seo, N., and Mendelsohn, R. (2008) A Ricardian Analysis of the Impact of Climate Change on South American Farms.*Chilean Journal of Agricultural Research*, **68**(1): 69–79.

Shue, H. (1992). The Unavoidability of Justice. In *International Politics of the Environment: Actors Interests and Institutions*, ed. *A.* Hurrel and B. Kingsbury. Oxford: Clarendon Press, pp. 373–97.

Shue, H. (1993). Subsistence Emissions and Luxury Emissions. *Law & Policy*, 15(1), 39–60.

Shue, H. (1999). Global Environment and International Inequality. *International Affairs*, 75, 531–45.

Smets, H. (1994). The Polluter Pays Principle in the Early 1990s. In *Environment after Rio: International Law and Economics*, ed. L. Campiglio, L. Pineschi and D. Siniscalo. London: Graham & Trotman/Martinus Nijhoff, pp. 186–87.

Smith, A. (1759). *Theory of Moral Sentiment*, London: A. Miller.

Smith, P. (2007). Climate Change, Mass Migration and the Military Response. *Orbis,* 51 (4), 617–33.

Stavins, R. (2014). Can There be a Positive Prognosis for Climate Negotiations? March. URL: www.robertstavinsblog/2014/03/. Accessed May 12, 2014.

Steininger, K.W., Lininger, C.,Meyer, L.H., Munoz, P., and Schinko, T. (2015). Multiple Carbon Accounting to Support Just and Effective Climate Policies. *Nature Climate Change*.

Stern, N. (2009). *The Global Deal: Climate Change and the Creation of a New Era of Progress*.New York: Public Affairs.

Stiglitz, J. (2012). *The Price of Inequality: How Today's Divided World Endangers Our Future*, New York: W.W. Norton.

Torvanger, A.&Godal, O. (2004). An Evaluation of Pre-Kyoto Differentiation Proposals for National Greenhouse Gas Abatement Targets, International Environmental Agreements: Politics. *Law and Economics* **4**(1): 65–91.

Tutu, D. (2007). We Do Not Need Climate Change Apartheid in Adaptation. In*Human Development Report 2007/2008: Fighting Climate Change: Human Solidarity in a Divided World*, ed. United Nations Development Program. New York: UNDP, pp. 166–186.

United Nations(1992). United Nations Framework Convention on Climate Change. Intergovernmental Negotiating Committee for a Framework Convention on Climate Change.

United Nations Conference on Trade and Development (UNCTAD) (2012). *Trade and Development Report*. Geneva.

United Nations Framework Convention on Climate Change (UNFCCC) (1997). "Proposed Elements of a Protocol to the United Nations Framework Convention on Climate Change." URL: http://unfccc.int/cop5/resource/docs/1997/agbm/misc01a3.htm. Accessed June 8, 2013.

Vanderheiden, S. (2008). *Atmospheric Justice: A Political Theory of Climate Change*. Oxford: Oxford University Press.

Vanderheiden, S. (2011). Globalizing Responsibility for Climate Change. *Ethics and International Affairs*, 25(1), 66–84.

Verheyen, R. (2005). *Climate Change Damage and International Law: Prevention, Duties and State Responsibility*. Leiden: Brill.

Verheyen, R., and Roderick, P.2008: *Beyond Adaptation: The Legal Duty to Pay Compensation for Climate Change Damage*. London: WWf-UK.

Volger, J. (1995). *The Global Commons: A Regime Analysis*. Chichester, UK: John Wiley & Sons.

US Department of Defense (USDoD) (2010). *Quadrennial Defense Review*, Washington, DC: US Department of State.

UNCLOS III (UN Law of the Sea III, 1983). Article 194.2.

Wang, J., Mendelsohn, R., Dinar, A., Huang, J., Rozelle, S., and Zhang, L. (2008) '*Can China Continue Feeding Itself? The Impact of Climate Change on Agriculture'. W*, World Bank Policy Research Working Paper 4470. Washington, DC: The World Bank.

Warner, K. (2010). Global Environmental Change and Migration: Governance Challenge. *Global Environmental Change*,20, 402–13.

World Bank et al. (2003). *Poverty and Climate Change: Reducing the Vulnerability of the Poor through Adaptation*. Washington, DC: World Bank Group.

World Bank (2010). *World Development Report 2010: Development and climate change*. Washington, DC., World Bank.

World Bank (2012). *Turn Down the Heat – Why a 4°C Warmer World Must Be Avoided*. Washington, DC: World Bank.

World Commission on Environment and Development (WCED). *Our Common Future*. Oxford: Oxford University Press, 1987.

Wright, B.D. (2011). The Economics of Grain Price Volatility. *Applied Economic Perspectives and Policy*, **33**(1), 32–58.

Young, O. (2010) *Institutional Dynamics: Emerging Patterns in International Environmental Governance*, Cambridge, MA: The MIT Press.

Index

CPSIA information can be obtained
at www.ICGtesting.com
Printed in the USA
LVOW09*1055310518
579107LV00007B/64/P